A SOCIAL ATLAS OF LONDON

JOHN SHEPHERD / JOHN WESTAWAY / TREVOR LEE

CLARENDON PRESS / OXFORD 1974

Oxford University Press, Ely House, London W.1

GLASGOW NEW YORK TORONTO MELBOURNE WELLINGTON
CAPE TOWN IBADAN NAIROBI DAR ES SALAAM LUSAKA ADDIS ABABA
DELHI BOMBAY CALCUTTA MADRAS KARACHI LAHORE DACCA
KUALA LUMPUR SINGAPORE HONG KONG TOKYO

PAPERBACK ISBN 0 19 874030 1

CASEBOUND ISBN 0 19 874026 3

PRINTED IN GREAT BRITAIN BY
COMPTON PRINTING LTD, AYLESBURY, BUCKS

Contents

Acknowledgements

One of our hopes for this atlas is that it will bring a knowledge of the internal structure of London to a much wider audience than has so far been the case. If it does, then we have mostly the Department of Geography at the London School of Economics and the School itself to thank. The Department of Geography has been a most helpful and congenial unit in which to work, while the School has provided the resources of personnel and facilities without which the atlas could not have been produced in its present form.

Most of the maps in the atlas have been produced by means of computer programs devised by Margaret Jeffery of the Department of Geography and written by Hazel O'Hare of the Computer Unit at the L.S.E. We feel that their work will be recognised as an important contribution to the presentation and analysis of census and other area-based information. In the planning stages of the atlas Richard Rieser gave us much useful advice and compiled the statistics for a number of maps. Professor Emrys Jones, whose *Atlas of London* is now a standard work in the field of urban cartography, was throughout a source of encouragement and constructive criticism. Janet Baker and Ann Stanton of the Drawing Office drafted and finished several of the hand-made maps for us, and Derek Summers braved two raw November days to take the photographs for the cover. To all of these people, to Janet Fox and the Departmental secretaries who typed the commentaries, and to Helen O'Hare and Robert Walker who helped to prepare the computer maps, we extend many thanks. It has been a pleasure to work with them all.

Finally, we take this opportunity to acknowledge the work of all those people, known and unknown to us, at the G.L.C., the I.L.E.A., the London Boroughs and the Office of Population Censuses and Surveys, who collect and tabulate the information which is the basic material for the study of London's social and economic structure.

John Shepherd, John Westaway,
Trevor Lee

London School of Economics, March 1974

Introduction : the spatial structure of London

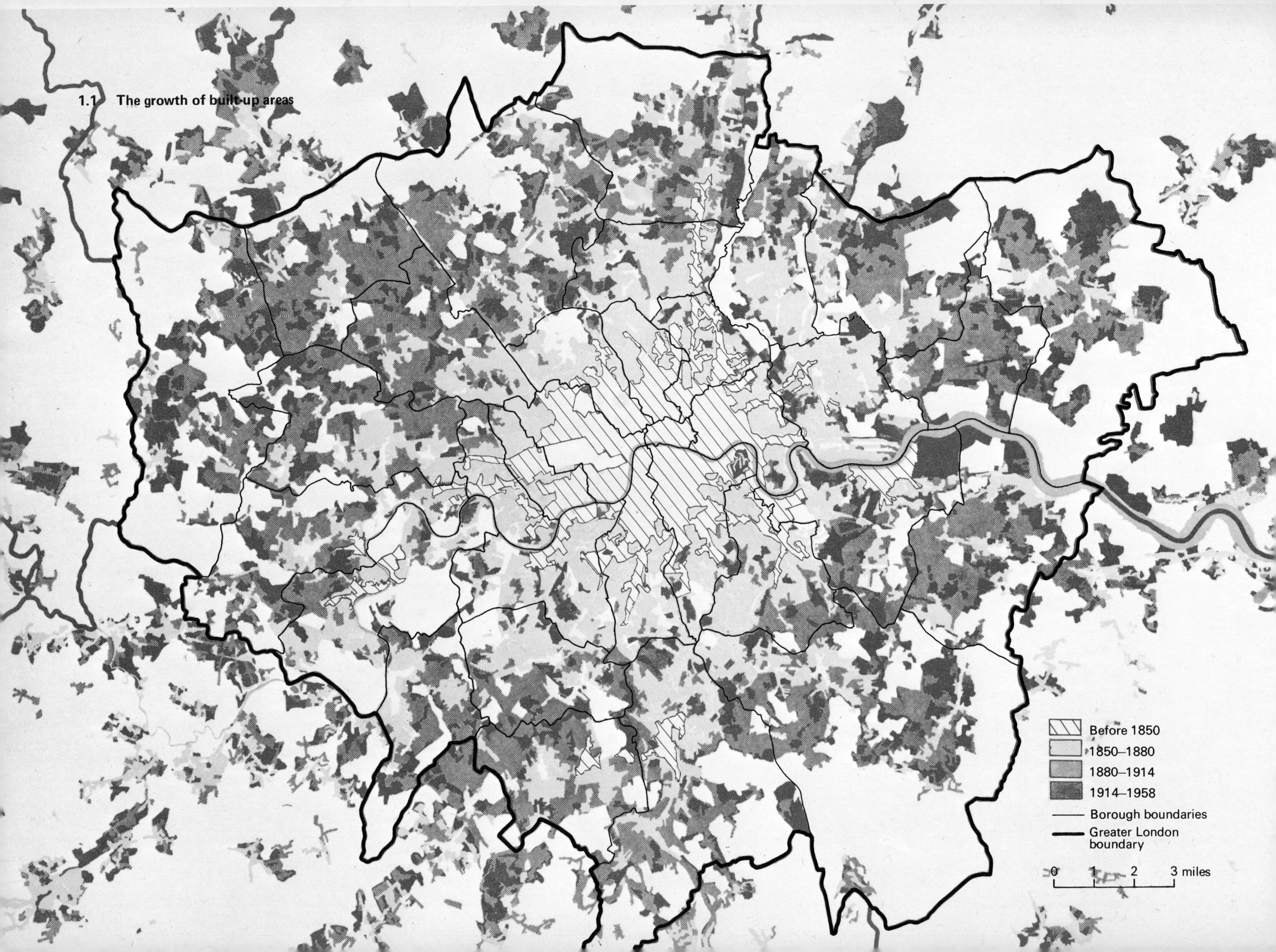

1.1 The growth of built-up areas

Introduction: the spatial structure of London

Simplifying the city

This atlas is a simplified representation of the huge and complicated social and economic system known as Greater London. The need for simplification arises as soon as we attempt to look at any city as a whole, for although we may feel that we have grown to understand the way our own neighbourhoods are laid out and the sorts of people who live in them, we will almost certainly have much less detailed knowledge of adjacent areas and probably very little idea at all of those areas on the other side of the city; and our understanding of broader, external forces affecting life in our neighbourhood may be slight. How very much more difficult it is to grasp the social geography of a world metropolis like Greater London, with a built-up area extending for fifteen miles in all directions around the centre, a population of more than seven millions, and a history of two thousand years.

There are a number of practical reasons for attempting to learn more about the city. During the past two or three decades, and in London particularly since the advent of the Greater London Council in 1965, large numbers of experts in urban affairs, including town planners, geographers, sociologists, architects, and engineers, have made valuable contributions to a deeper understanding of the city. Yet, to workers in other fields and to the interested or involved citizen, the reports on this work are often difficult and costly to obtain and are rarely written in a language or style that are intelligible to the non-specialist. Moreover, there is no simple and easily available exposition of the overall context, no 'birds-eye view' of the many different aspects of Greater London against which the more detailed research and conclusions might be compared.

The need for a wider appreciation of this more general framework is especially important in the light of recent developments in the way our cities are planned. In London planning takes place on two levels. The Greater London Council deals with matters affecting London as a whole and the London boroughs concern themselves with more local issues. The two-tier division of planning powers has now been extended to other metropolitan areas in England. It is therefore necessary for the general public, which is being encouraged to participate more fully in the planning process, to grasp the meaning of terms like 'structure' and 'strategic' plans; to understand how they relate to 'local' and 'action area' plans; and to know how the advantages and disadvantages of planning decisions are assessed and compared both for the city as a whole and for the localities most directly affected.

In the field of social welfare it is also necessary to have some knowledge of the structure of the total physical and social environment of the city. Poverty is no longer looked upon as simply a lack of money income, although that is obviously the single most important ingredient. It is also seen as a lack of *access*, defined either as physical access to or knowledge of decent housing, jobs, good schools, and leisure facilities. We now know that the poor in the city are likely to be deprived of most or all of these things at once. Hence the importance of the question 'where is the poverty in the city', and our concern to define the areas of *multiple deprivation* in the city and to develop policies of 'positive discrimination' in the allocation of extra resources to such areas.[1]

Our approach to simplifying Greater London is as follows. In this introduction we first briefly describe how information on the social, economic, and political life of a city can be conveyed in the form of a map. Yet a map is very much more than a convenient method for presenting what might otherwise be a rather arid collection of statistics. Its property of organizing information geographically or spatially means that it can be regarded as a model of one detailed aspect of the city, and the regularities or patterns that it reveals are of considerable value to an understanding of how those patterns came into being. We therefore move from a description of how the maps in this atlas were compiled to an outline of three of the many theoretical frameworks for the internal organization of cities, each suggested by geographical patterns apparent on maps.

Each of these three frameworks is concerned with explaining one limited part of the total city mechanism. The first considers the arrangement of centres of retail and other service activities within the city, while the second describes the housing and social patterns that result from certain conditions of expansion of an urban area. The third relates specifically to Greater London and links the distribution of different socio-economic groups to the nature of the underlying topography.

[1] Among the programmes of positive discrimination in social welfare which include a strong geographical element are those for *educational priority areas* (announced in a D.E.S. Circular on the School Building Programme, 1967); *community development projects* (Urban Aid Programme, 1969), and *housing action areas* (White Paper, Better Homes: The Next Priorities, June 1973). In addition, local authority social workers are now deployed in integrated teams on an area basis under the Local Authority Social Services Act, 1970.

If we wish to go beyond these three theories and attempt to link a wider range of our maps of Greater London together then our theory about urban patterns must be correspondingly more generalized. In addition, we must be prepared to treat more aspects of the city as 'exceptions' to the theory, even though they may fit well into the scheme of other, more specific theories. As a background to the structure of Greater London, we therefore introduce four key maps—the growth of the built-up area, the distribution of major land uses, the communications network, and the population density— and view them in the light of the concepts of accessibility, competition for land, and improvements in transportation. Together, these concepts suggest a theory of urban land values which has proved useful in clarifying the nature of the relationships between the built environment of the city and its human environment as represented by the distribution of its population. We conclude the introduction by distinguishing some of the unique features of the geography of Greater London that are not accounted for by the working of these theoretical principles alone.

The maps which then follow in the atlas are grouped together in a number of distinct sections, each of which can be seen as a major component part of London's physical and social structure. Individual maps may take the form of a *line drawing* of streets, administrative boundaries, and other outlines of the city; they may be shaded or *choropleth* maps built up from information on very small sub-areas of the city; or they may convey information in the form of *symbols* like pie diagrams or population pyramids.

Most of the maps are choropleth maps and have been constructed on the basis of the electoral wards of London, using information derived from the Census of 1971 and other recent sources. There are over 650 wards in Greater London, the majority of them of a territorial size that an individual might be acquainted with, even though he or she might not know the precise delimitation of ward boundaries.[1]

In order to produce a choropleth map social characteristics like population density, housing conditions, or social class are measured for each ward. Each ward's score for all the individuals (physical objects or people) located in it characterizes the ward itself. The process of generalizing what began as a great mass of information on individuals is carried further by ranking on a scale all the wards according to their average measure, grouping them into a convenient number of classes or intervals (usually five or six), and then shading each ward on the map according to the class-interval into which it falls.[2]

A choropleth map gives us a picture of an aspect of the city that has been built up from information for very small sub-areas within it. It enables us to compare the characteristics of individual wards much more easily than we could by reading the original tables of census statistics, and it also helps to establish the extent to which certain sorts of people, activities, and environmental conditions are concentrated in particular parts of the city. The process of recognizing the shape and intensity of geographical or spatial patterns represents one useful method of generalizing phenomena which, in turn, makes it possible to compare and contrast the patterns on two, or perhaps more maps simultaneously. In other words, it enables us to make some very rough, visual *correlations* between the characteristics displayed on the maps. The next stage in understanding the city is to explain why these relationships exist and how one characteristic of the life of the city affects another.

City patterns

The first of the three frameworks referred to above derives from the fact that London is the largest single city in Britain while at the same time it embraces a number of distinct centres of activity. In geographer's language London is at the head of the national 'hierarchy of central places' or towns and also contains its own hierarchy of centres. The essence of this idea was once described by Marghanita Laski in a more imaginative way when she spoke of London as providing simultaneously the facilities of a 'metropolis, town, and village'.

London is the premier city of Britain, both in terms of the absolute size of its population and in the range of service functions that it performs. Within central London are Parliament and the national government, the major museums and galleries, the head offices of national and international companies, and some of the world's most prestigious and specialized shopping streets. A huge 'catchment area' population, extending for some services well beyond the boundary of Greater London, is necessary to support, for example, the concentration of major department stores along Oxford Street, of theatres in the West End, and of fashionable clothing shops along the King's Road, Chelsea.

[1] The ward outline map in Appendix 3, together with the list of ward names, will help in locating a particular ward, or group of wards of interest. The data for the City of London, which is administratively and demographically unique, has usually not been mapped but the area is marked CITY.

[2] A detailed description of the method of choropleth mapping and a guide to interpreting such maps drawn at different geographic scales or levels of aggregation of areal units can be found in Appendix 1.

1.2 Major land uses

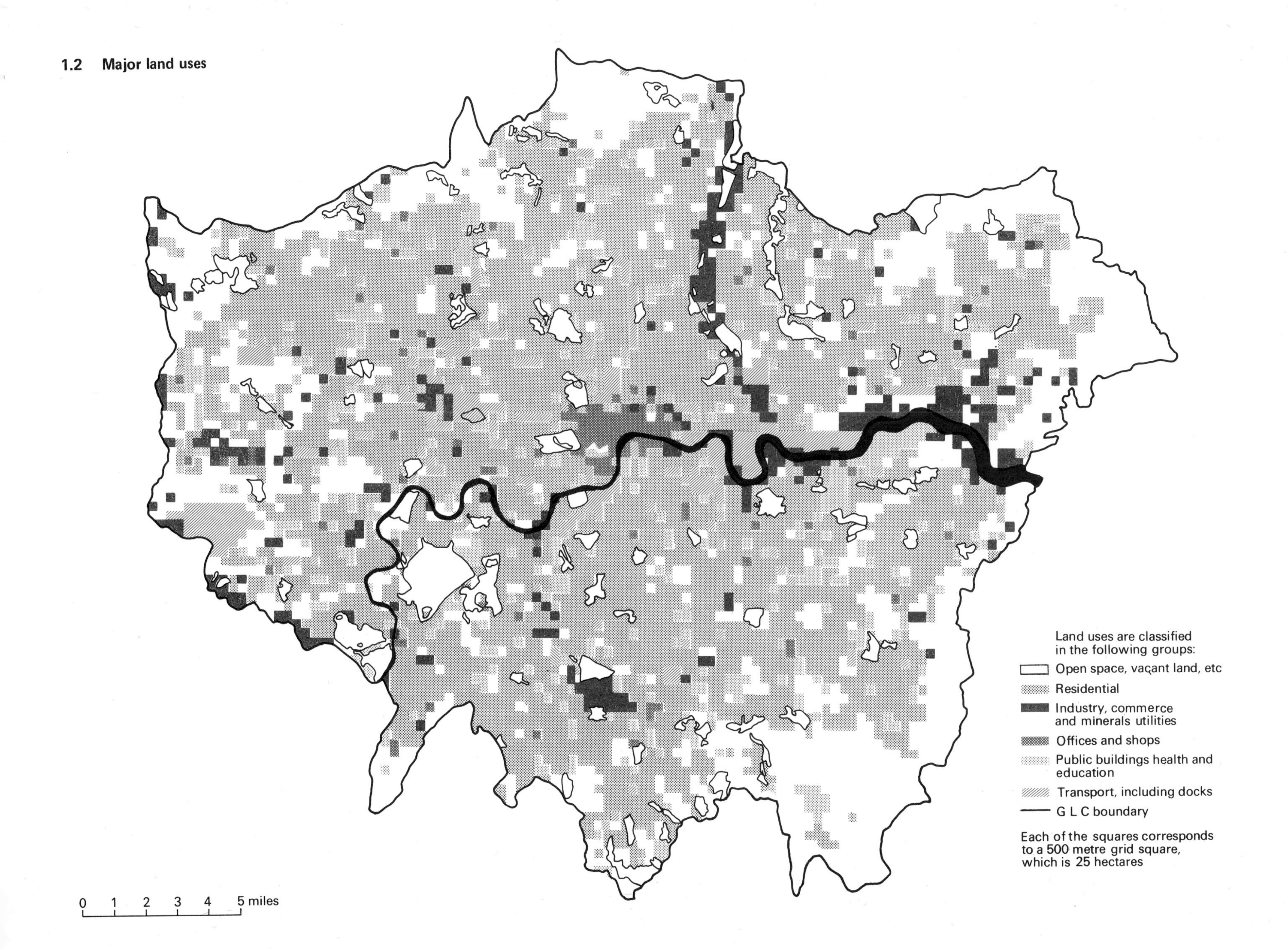

Yet that part of the catchment area for the centre which lies within Greater London also contains several 'towns' comparable in size and function to towns in the provinces. Places like Ilford, Lewisham, Brixton, and Harrow, for example, have their own large shopping centres, civic amenities, and local newspapers, and they generate their own form of community consciousness.

Finally, at the lower levels of the hierarchy of places within London, are the 'villages', which themselves range in size from a recognizable retail centre to a small parade of shops or a corner store. They have all been long since enveloped by the spread of building but, in some, the village atmosphere is deliberately cultivated for social and commercial reasons as in Hampstead, Highgate, and Blackheath. Others, like Camberwell with its remnant of a village green, and the Angel, Islington with a wide market street, retain physical vestiges of their origins but are not as well cared for. Still others, like Haggerston, Nunhead, or Northfields, may be little more than names to the outsider, though they are metropolitan villages in the consciousness of those who live there.

The second framework for describing the internal organization of cities was suggested by the American sociologist E.W. Burgess. Burgess's 'concentric zone' theory was developed in the context of the inter-war growth of cities in the U.S.A., particularly Chicago. At the heart of the city lay the central business area. Surrounding it was a 'zone of transition', an area of deteriorating tenement blocks and factories, the home of the most recent arrivals to the city or those who could not escape to the suburbs. Next came three zones of successively better residences and more stable social structure: a zone of single- and two-family dwellings inhabited by the 'respectable' working classes and second generation immigrants; a zone of exclusive middle-class suburbs; and a commuting zone.

London and its surrounding region is often described in terms of a number of concentric zones and there are several officially-defined areas, which are useful for this purpose. Central London, for example, has been defined by the Office of Population Censuses and Surveys as corresponding roughly to an area bounded by the line joining the main railway terminals at London Bridge, Waterloo, Victoria, Paddington, King's Cross, and Fenchurch Street; while the term *Inner London* commonly refers to the area covered by the Inner London Education Authority (Map 2.6), formerly the area of the London County Council. Including, as it does, most of Greater London built before 1914 (Map 1.1), it contains much of London's stock of obsolescent dwellings, though it is by no means a territory of uniformly bad housing.[1]

The next zone of development is usually described as Outer London and consists of the inter-war suburbs which are wholly enclosed by the boundary of the Greater London Council. Then, stretching for twenty or thirty miles beyond the green belt, is the Outer Metropolitan Area (Map 2.6), defined by central government planners in the mid 1950s on the basis of local government areas experiencing rapid population growth and increasing levels of commuting to London.

Greater London, therefore, can be, and often is, described according to a pattern of concentric zones, but this does not mean that such a description is necessarily completely accurate. Neither does it mean that the processes by which such a pattern developed are the same as those of the Burgess theory. London grew over a very much longer period than Chicago and in a very different social and political context. Although there are marked contrasts between Inner and Outer London in a wide range of characteristics, there are also many detailed departures from a purely zonal pattern which offer additional insights into the process of internal differentiation of London in particular and cities in general.

A completely different framework particular to London's social geography has recently been proposed by Peter Willmott and Michael Young, who analysed social-class maps based on census information for 1966. They identify three main elements in the pattern—a central residential and business district, an extensive cross-shaped area into which the majority of the working class is concentrated, and four suburban wedges occupied by the middle classes. The central residential district lies to the west of the West End and City of London in the boroughs of Westminster and Kensington and Chelsea, and is mainly upper-class. The working-class 'cross' extends east to west from Uxbridge to Dagenham, north along the Lea Valley to Enfield, and southwards along the River Wandle to Morden and Croydon.

According to Willmott and Young the explanation for this pattern lies in the distribution of low-lying ground associated with river valleys and the suitability of such land for docks, industry, and communications. The more favoured areas of London outside the centre are therefore on the higher ground, forming the suburban wedges of the north-east

[1] The London Borough of Newham (formerly the county boroughs of West and East Ham), is often grouped with the 'official' Inner London boroughs on the grounds that it is similar to them in age of development and in its social and environmental problems.

1.3 Transportation network

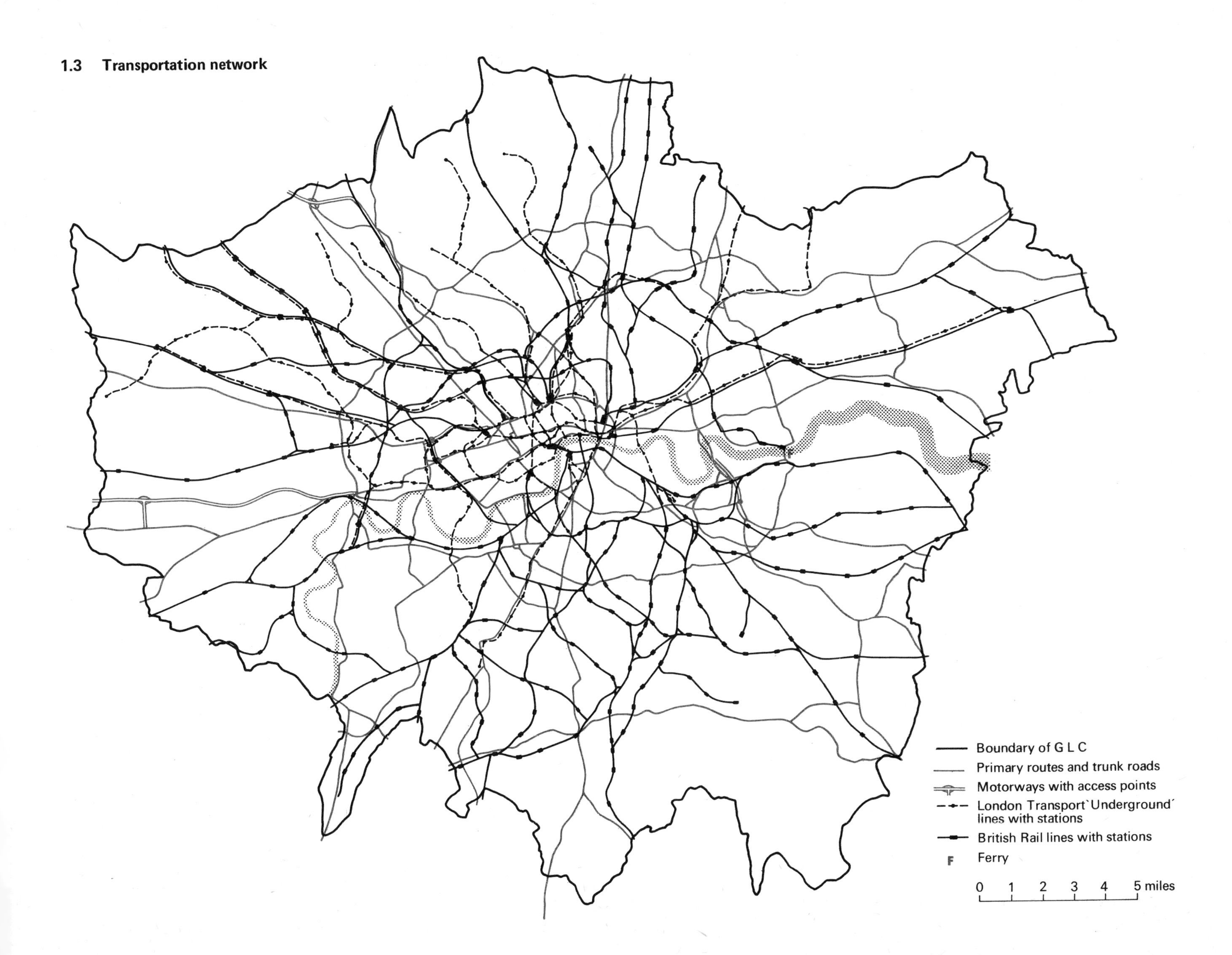

(Waltham Forest and Redbridge), the south-east (Bromley), the south-west (Richmond and Kingston), and the north-west (Harrow and Barnet).

There is little doubt that the lie of the land has affected the character and growth of London, notably through the distribution of parks and open spaces, the construction of the main drainage system in the nineteenth century, and through the location of roads and railways. It may also be that topography is a factor in the social character of a particular neighbourhood but it is doubtful if it can explain fully the pattern for London as a whole. Given the size and complexity of London we cannot ignore the much wider social and environmental forces which are at work, among the most important of which are inequalities in the distribution of wealth, the quality of the physical environment, and access to opportunities for personal development. Too much emphasis on physical geography as a determining factor in the social patterns of the city could lead to the impression that we are powerless to control our urban environment. Greater London, like any city, is a man-made phenomenon and in the course of its development numerous topographical obstacles have been surmounted. Today its physical environment is largely what we have made and are making it. If that environment has important social implications, and research is showing that it has, then we are ultimately responsible for them as well.

The structure of London

As previously noted, four maps have been chosen to illustrate the major structural features of the physical and human environment of Greater London. These are the *growth of the built-up area* (Map 1.1), which indicates the overall shape, direction, and extent of urban growth during different time periods; the distribution of *major land uses* (Map 1.2), showing the current patterns and geographical relationships among the important land-using activities in London; the *transportation network* (Map 1.3), indicating the patterns and relationships among the conveyors of people and goods—road, surface rail, and underground; *population density* (Map 1.4), which shows where people live in London and which is one representation of intensity of land-use.

It is useful to begin the task of describing and relating these introductory maps to each other by thinking of London as a huge system of resources including land, capital invested in offices, factories, houses and transport facilities, and labour. Land and capital have fixed locations. Labour is relatively mobile. The relationship between the fixed and the mobile resources, the ease with which physical resources can be brought together, and the ease with which people can move from home to workplace and between workplaces, all this can be called *accessibility* and should be considered as another important resource of the city. Accessibility is, moreover, a special sort of resource since it helps to determine the availability of other resources. It is reflected in the cost of land, capital, and the movement of people and goods. Changes in accessibility brought about by advances in methods of transport and investment in facilities for transport can radically alter the availability and cost of the other resources and, through them, the form and structure of the city.

Accessibility explains the most important factor common to all four structure maps of London, namely the dominance of the centre. Examination of these maps will show that the general pattern of growth has been outwards from the centre; service employment in offices and shops is concentrated at the centre; the highest population densities are clustered in a ring around the centre; and the main lines of communication converge on the centre. The centre of a city is the area that is most accessible to all other parts of the city. Within the central area the different activities are clustered closely together, permitting fairly easy and frequent contact between them. It is interesting to note in this respect that, despite the introduction of the telephone and other means of electronic communication, the central area of the city is still highly valued for the ease of personal contact which it provides.

A simple extension of this notion of the accessibility value of the central area helps to explain other aspects of the structure of the city. At the centre, land has a very high monetary value because it is most accessible to all parts of the city. Only those activities which can afford to pay very high rents are able to locate at the centre. Hence the concentration of offices and shops at the centre of the land-use map of Greater London; no less than 60 per cent of all London's office floor space is located in the central area, which represents less than 2 per cent of the total land area of Greater London. These activities can 'bid-up' the rent and keep out or force out other land-users like industry and housing.

Related principles can be used to explain the overall pattern on the population density map. In our economic system housing competes with other activities in the city for a share of the available land. We have seen that the money value of land is high enough at the centre to force out most non-business activities. This produces a characteristic 'hole' in the population density map, representing a relative lack of residential density at the centre. There are, of course, people who live in the centre of London. Some, the extremely rich, can afford spacious flats and houses in the heart of the city. Others rent tiny flats or perhaps basement bed-sitters, or live in local authority or charitable trust tenements built in the last century. Population densities are extremely high around the 'hole' at the centre because most people have no choice but to 'trade off' advantages in accessibility to the centre for a relatively small amount of living space. Many people pay relatively high rents for restricted living space but, at the same time, save money on the cost of transport to work. There appears, therefore, to be a paradox in the social geography of the city whereby some of the poorest people live on the most expensive land. The paradox is explicable, however, when we realize that the high rent of this land is shared among many more people.

Finally, surrounding the very high-density zone are the generally much lower-density suburbs. Here, most people pay relatively less for spacious housing but have to pay substantial travelling costs if they work in the West End or the City.

The overall pattern of population densities can also be related to the 'age of built-up area' map which is, in turn, a product of developments in the technology of mass transportation and hence of accessibility. Before 1850 London was predominantly a pedestrian city and the furthest extent of the built-up area was limited by the three to four miles covered in an hour's walk to work or a half-hour in a horse-drawn carriage. The suburbs of the mid-nineteenth century were places like Camberwell, Clapham, Holloway, and Highbury, and similar areas of large houses, which have since been divided into smaller flats and bed-sitters to accommodate many more people.

The steam age in London began in 1836 with the opening of the Greenwich railway to London Bridge and, by the time the era of railway expansion had run its course (roughly 1905–10), the suburbs of middle-class 'villas' and working-class terraces had followed the lines out to Willesden in the west, northwards to Tottenham and Enfield, eastwards to Ilford and Becontree, and southwards to Penge and Croydon. However, it was the Underground and, later, the private car, which were responsible for the most intensive period of suburbanization in London's history. The tubes were cleaner and faster than the steam trains and the extensions of the Bakerloo line as far as Watford in 1917, the Northern line to Edgware and Morden in the early 1920s, and the Piccadilly line to Cockfosters and Hounslow thrust the limits of urbanization still further outwards. Sometimes tube stations were built in the fields of London's countryside, anticipating the residential growth to follow. As a consequence, in the period 1919 to 1939, the built-up area of London more than tripled in extent—a situation which ultimately led to the imposition of planning controls to halt the further spread of the metropolis. The motor car and bus, with their greater flexibility of movement, then made possible the infilling of the area between the tentacles of growth along the rail lines.

The analysis so far has used the concepts of accessibility and the relative values of accessibility to different land-users to produce an extremely simplified 'model' of London's basic structure. As we have seen, it appears to fit the general pattern illustrated by the four maps tolerably well. Yet in all the maps it is possible to discern several departures from a simplistic focus on a single, all-powerful city centre exercising its influence equally in all directions. We may consider these departures under four headings: employment, sub-centres, historical associations, and planning.

As shown on the land-use map the working areas of Greater London consist of three main elements. There is the massive concentration of offices and shops at the centre, which we have already discussed, together with a number of very much smaller centres of service employment in other parts of London. There are large areas dominated by transport industries like the Heathrow Airport complex and the Thames-side docklands, the latter currently undergoing great changes as the locational and amenity potential of the riverside close to central London is re-assessed and the land re-developed. Finally, there are the five main industrial areas of Greater London in lower Thames-side, west London, the Lea Valley, the Wandle Valley, and the belt of manufacturing on the northern and eastern periphery of the West End and the City.

Four of these five industrial work areas are evident on the map as the predominant category of land-use in their respective locality. Their location has been determined by lines of communication to the major industrial areas in the North and in the Midlands, and the local availability of suitable building land. Lower Thames-side in the east and the more scattered west London areas were mostly established between 1920 and 1939 to manufacture consumer appliances, vehicles, and scientific goods, usually in relatively large factories on a production line basis and employing large numbers of semi-skilled workers. Hoover at Perivale and Ford's of Dagenham are among the best known of these concerns. The other two extensive industrial areas follow the lower-lying, marshy ground in tributary valleys of the River Thames. On the north side the factories of the Lea Valley stretch from Poplar to Enfield while those to the south, in the Wandle Valley, are mainly clustered in Merton and Mitcham but also extend as far as Wandsworth. The lower reaches of both valleys still contain some of the original industries of the area—the production of food and drink, chemicals, and soap and candle-making—a number of them having moved from the central London area in the late nineteenth century. But, as in west London, many new establishments began production further up each valley in the period between the two world wars. Many of them were situated in factory estates erected by speculative builders along the arterial roads of the 1920s, like the North Circular and the Great Cambridge Road.

London's fifth industrial area is centrally located but is not readily apparent on the sort of land-use map we have here. This fact emphasizes its special character as a central-city, and therefore highly intensive, land-user. The factories and workshops of this central London zone are typically very small; indeed, several may occupy a single building. As a simple land-use category, therefore, industry in this zone rarely predominates. Its typical trades include clothing, furniture-making, printing, metal-working, jewellery-making, and precision engineering, often located in distinct 'quarters' between Camden Town and Oxford Street in the west and Stepney Green and Limehouse in the east. Because of the extremely intensive character of manufacturing in this area it can easily be forgotten that there is a manual work-force of over a quarter of a million people, mostly highly skilled and experienced, in central London and its immediate environs.

Turning now to the major subsidiary centres of Greater London we can recognize a number of them in different ways on one or more of the four structure maps. They may appear as nuclei of early urban development surrounded by more recent growth, like Woolwich or Richmond; they may be local points of convergence in the London communication network; or they may be one of the numerous centres of either offices and public buildings or of relatively higher-population density, like Wandsworth, Barnet, and Bromley. Two such sub-centres — Kingston-upon-Thames and Croydon—are apparent on all four structure maps, and are of particular interest in that they illustrate the contribution of administrative functions and political decision-making to the development of the city.

Kingston is now the main centre for the London borough of Kingston-upon-Thames. It is therefore within the Greater London area, and yet it retains its functions as the county town and administrative headquarters of the County of Surrey. Croydon, on the other hand, was a county borough prior to the reform of London's local government in 1965. As such it had the powers to draw up and implement its own development plan and, in the late 1950s, the council took the decision to create a major suburban office centre in Croydon. The town was able to take advantage of its good transport links both with central London and the surrounding residential areas and to offer competitive office rents to attract firms which might otherwise have sought locations in the City or the West End. Since then, other suburban centres have followed this example, though none quite so spectacularly as Croydon.

Thirdly, London's historical associations are for many people, resident and visitor, the very essence of its unique character. Few cities, for example, have such a rich heritage of beautiful parks within or very close to their central areas. Yet the Royal Parks of London like St. James's, Regent's Park, and Greenwich Park were once the exclusive preserve of monarchs and nobility. Now they provide peace and respite from the noise and dirt of city streets for a mass of people. Parks, commons, and other open spaces in the centre of London and elsewhere also lend prestige and desirability to the residential areas around them—a factor which undoubtedly helps to mould the social-class character of certain neighbourhoods. In contrast other parts of inner London are still without sufficient open space, despite the stated intentions of numerous town-planning documents. Of the inner London boroughs Wandsworth and Westminster each have over 1000 acres of public open space within their boundaries, while Islington has only 78. It is important, therefore, to keep the historical legacy of the city in mind, including the character and distribution of its ancient buildings and street patterns, when attempting to describe it by a simplified model of its structure.

Finally, in comparing a real city to an idealized distribution of land uses or demographic patterns such as those described by the urban rent/accessibility model, it is necessary to take into account the distorting effects of public authority intervention in housing, employment, transport,

1.4 Population density

and town planning.

Certain physical effects of town planning measures are clearly evident on a number of maps in the atlas. Strict controls on development in the Metropolitan Green Belt (Map 2.6) have effectively halted the outward growth of London at the stage it had reached by 1939 (Map 1.1) In housing the predominant pattern of owner-occupied dwellings in the outer suburbs is interrupted in places by council estates of subsidized housing, which may be 'out county' estates built by the L.C.C. to rehouse people from the slums of inner London, or may have been undertaken by local housing authorities in the suburbs (Maps 3.2 and 3.3); within inner London, council housing usually provides a much higher standard of accomodation than adjacent privately rented dwellings (Maps 3.5a and b). Town centre redevelopment schemes planned by local authorities and approved by central government have determined the location and growth of service employment in the outer suburbs and the decline of manufacturing employment in inner London has been partly due to industrial location policies carried out in the *national* interest (Maps 1.2 and 6.1). Finally, the commentaries to the maps of bus and rail accessibility in London (Maps 6.4b and c) point out some of the disparities in the provision of public transport services. Without the unified control of public transport that emerged for London as a whole in the 1930s, these inequalities in accessibility might have been considerably greater than they are today.

In this introduction we have described our approach to compiling a social atlas of London and have outlined a number of ways of interpreting the maps. The maps that follow have several uses. As individual works of reference they illustrate the spatial distribution of particular phenomena; in groups they can provide a composite description of a particular locality or sector of London; and, taken as a whole, they should convey a picture of London as a metropolis with unique characteristics as well as features which conform to our idea of the structure of the Western city. At whatever level it is studied, the atlas is a geographer's view of London and its people which provides new insights into the social environment of those who live and work there.

The social development of London

2.1 Tudor London

Braun and Hogenburg's map of London, published in 1574, is significant for its representation of the whole 'metropolis' of London. It shows not only the walled City of London but also Westminster, joined to the City by the palaces and grand houses along the Strand, and Southwark and the surrounding 'out-parishes' and villages that were to grow so rapidly in the succeeding century.

The institutional forces which helped shape London's growth are clearly depicted by the map. London was developing as a dual-centred capital. The City was the great commercial and manufacturing centre, virtually independent in administration and in the regulation of business enterprise, while Westminster was the royal and ecclesiastical city and the centre of an emergent national government. To the east, however, the Tower of London, military outpost of the Crown, served as an uncomfortable reminder to the City Fathers of the ultimate powers of the sovereign state.

Scattered around the periphery of the City and Westminster were the newer and, for the most part, uncontrolled suburbs. In only forty years from 1560 to 1600 the population of London had more than doubled from about 90,000 to 200,000 inhabitants, mainly because the rate of in-migration was even higher than the very high death rate among natives of the city. Migrants came from abroad and from the provinces and crowded into hovels along the Fleet River, already an open sewer, in Clerkenwell, and in Smithfield. Others, notably the French, Dutch, and Jews set up in weaving, silk throwing, tailoring, and dyeing in Spitalfields, St. Giles, and Whitechapel. Eastwards, extending along the Thames from the Tower to Radcliff, a straggling community of seafarers, shipbuilders, and ship's chandlers was developing with the Tudor expansion in foreign trade. Across the river, Southwark, Bermondsey, and Lambeth were becoming crowded with artisans as industry began to move out of the City to escape its apprenticeship regulations and municipal taxation.

There were, of course, the more salubrious and fashionable districts as John Stow describes in his *Survey of London* (1598). Grays Inn Lane, for example, was 'with . . . many fayre houses builded, and lodgings for Gentlemen, Innes for travellers, and such like up almost . . . to St. Giles in the Fields'. In St. Botolph's parish, Billingsgate an influx of rich foreigners had caused a considerable increase in rents for 'the nearer they dwell to the water side, the more they give for houses' and London Bridge itself was 'replenished on both sides with large, fayre and beautifull buildings, inhabitants for the most part rich merchants . . .'. New development to the west of the City was mainly a phenomenon of the Stuart period though, by the end of Elizabeth's reign, town houses of aristocrats and lawyers were appearing along Drury Lane and Chancery Lane.

The social and political effects of London's growth had repercussions for City and Crown alike. The immunity from the law provided by the ancient Liberties and other privileged places like Blackfriars and St. Martin's le Grand, meant that they became concentrations of criminality, riot, and sedition and centres of overcrowding and disease, which spread into surrounding localities. Riot and disorder was most likely to break out in areas of deep poverty and in those dependent on economically unstable trades like weaving and tanning, or with large concentrations of foreigners and Roman Catholics.

One possible solution to these problems was the extension to the expanding suburbs of the powers of the magistrates. This approach foundered, however, on disagreements between the City and Westminster on the exact form of enlarged local government and the City's fear that absorption of the suburbs would weaken its trading monopolies and alter the social composition of its Court of Common Council. Instead of undertaking a thorough reform of London's government, the City and the Privy Council therefore agreed to try to limit London's growth and issued a proclamation in 1580 prohibiting the building of new houses or the conversion into tenements of existing ones within three miles of the gates of the City. But enforcement of the building laws still required proper administration and, in its absence, London's earliest 'green belt' planning failed to halt the growth of the city.

2.1 Braun and Hogenburg's map, 1572

Housing

3.1 Housing Tenure and Social Structure

Housing in London is a scarce and expensive commodity. This is because there are over 150,000 more households in London (2·65 million in 1971) than there are dwellings (2·50 million)—a deficit which stems from the loss of housing stock in London during World War II and from a trend towards smaller households.* Since 1939 the *population* of Greater London has declined steadily but the number of *households* has continued to rise. These increases have stabilized since 1961. The trend towards a larger number of small households (1 or 2 persons) can be partly explained by smaller family size, the high proportion of young and single people in London, and the growing population of old people who also wish to maintain separate households.

The housing shortage in London has a number of undesirable consequences. One of these is that it causes overcrowding, as certain households may be forced to live in accommodation which has too few rooms for their needs, their health, and their privacy. It also promotes multiple occupation, and some households must share basic amenities (such as cooking or washing facilities) with another family or household. Other households may 'live in' with another family, as in the case of young married couples who remain in their parents' home after their marriage. But the most serious consequence of the shortage of housing in London is that there are many people who are left completely *homeless*; in 1972 there were over 11,000 homeless families in London applying for temporary accommodation.

The repercussions of this shortage extend far beyond the sphere of housing itself. Many of London's essential services, such as public transport and schools, are beginning to be seriously disrupted because of staff shortages; employees have difficulty in finding satisfactory accommodation at a moderate price.

People who have bought the house in which they live, or are in the process of paying off the mortgage, are known as *owner-occupiers*. Alternatively, accommodation, either unfurnished or furnished, may be rented from a private landlord (the *privately rented* sector), or from a public housing authority controlled by the G.L.C., a borough, or a new town corporation (*council tenants*).

Most people in London rent their accommodation, although owner-occupiers form the largest *single* tenure group (40·4 per cent of all households in 1971). Council tenants form 24·9 per cent of London's households, while the corresponding figures for privately rented unfurnished accommodation and furnished lettings are 23·6 per cent and 10·5 per cent respectively. Over the period 1961—71 there has been proportional growth of council tenants and owner-occupiers, decline of private unfurnished tenancies, and stability of private furnished tenancies.

The proportion of households in each tenure group is not the same throughout all boroughs of Greater London, as Map 3.1 clearly shows. A marked division is apparent between inner London and the outer boroughs, where the proportion of owner-occupiers is much higher. Within inner London much of the privately rented accommodation is focused in the western sector (from Hammersmith to Islington), while eastern boroughs (such as Tower Hamlets) have a high proportion of council accommodation.

The spatial separation of tenure groups largely reflects the historical growth of London. Most dwellings in inner London were built before 1919, whereas the bulk of the housing stock in outer London was built in the interwar period (1919-39). This means that most of the housing problems (in terms of dwellings unfit for habitation and lacking amenities) are concentrated in inner London and, as most rented accommodation is in inner London, it follows that this housing sector, and the people in it, bear the brunt of London's housing stress. Furthermore, different groups within the population tend to be concentrated in specific tenure groups. For example, the sheer cost of housing ensures that owner-occupation is concentrated among the middle classes, while privately rented furnished accommodation is the tenure group most closely associated with housing the young, the old, the very poor, and the immigrant.

Social status, income, and stage in the life cycle all interact to affect the type of tenure group an individual is likely to be in. The tenure group is likely to affect (in broad terms) the location of a household's residence in London, and this locational factor is closely linked with the quality of urban environment. Patterns of housing tenure in the conurbation are therefore a key element in the social geography of London and provide a social link between the built environment (the physical structure of London) and the social environment (the social structure of London in its spatial context).

*A household may be defined as a group of people, whether related or not, who benefit from common housekeeping. or a person living alone who is responsible for his or her own meals.

3.1 Tenure of households

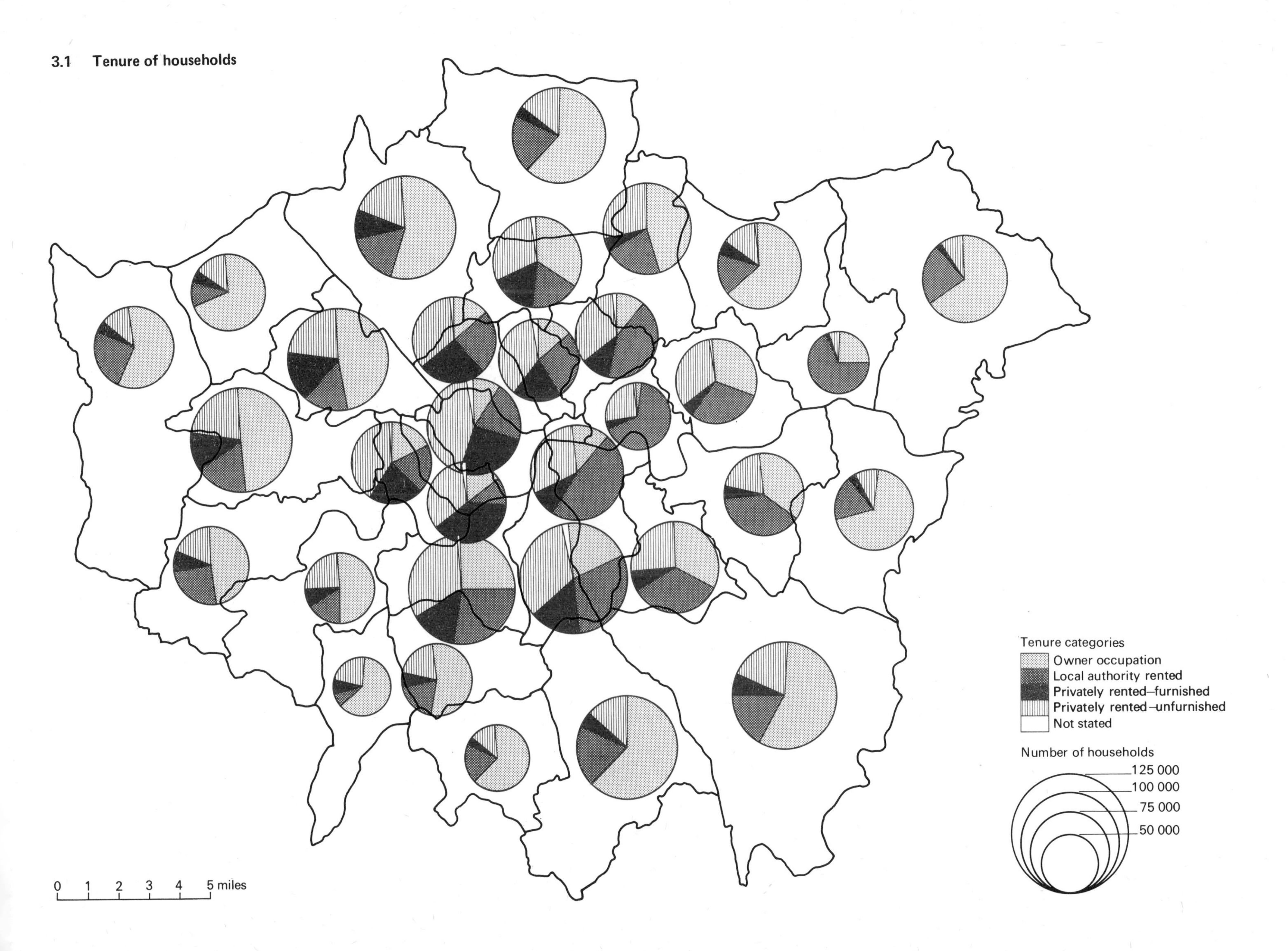

3.2 Owner-occupation

Owner-occupation is undoubtedly the most favoured sector of the housing market. Home ownership is widely regarded as an index of status and an object of personal aspiration. Moreover, it offers absolute security of tenure and mortgage repayments go towards the purchase of a capital asset which is likely to appreciate considerably in value. The financial incentives towards owner-occupation are also stimulated by tax relief on mortgage payments, which is widely regarded as a housing 'subsidy' comparable to that enjoyed by council tenants. Another financial paradox of owner-occupation is that the average weekly mortgage payments (including rates and water charges) for London's owner-occupants are considerably less than the average weekly rent for private tenants of furnished accommodation and only some 50p per week more expensive than those of local authority tenants (1971 figures). Once a mortgage is repaid, annual rates and water charges impose only minimal 'rent' on the owner-occupant.

There are more than a million owner-occupants in London, but as Map 3.2 shows, they form a far higher proportion of all tenure groups in the outer boroughs than they do in inner London. There are large areas of the inner city where fewer than one household in five is in owner-occupation, whereas in many of the peripheral areas more than four households in five fall into this group. The predominance of owner-occupants in the outer boroughs is broken occasionally by areas of decentralized public housing. This can be seen in the north-eastern tip of Havering (Gooshays ward) and throughout much of the borough of Barking.

As with all tenure groups, owner-occupation tends to be associated with specific social classes; the distribution of housing stock is closely interwoven with the residential segregation of social and economic groups in London. Owner-occupied households in outer London tend to be of medium size and the incidence of overcrowding is very low. Heads of owner-occupied households are predominantly non-manual workers and are concentrated in the higher income groups. Most of the houses in the outer boroughs were built in the interwar period, with the extension of rail services (predominantly underground services to the north and surface rail to the south of the Thames) providing the crucial links between place of residence and place of work. But, despite the decentralization of many job opportunities, accessibility in outer London is still relatively low (see Map 6.4).

The greater pressures for car ownership in families wishing to own a home in outlying districts further militate against the suburbanization of low-income families. Care should be taken, however, to avoid portraying outer London as a uniform area of high-quality housing. Pockets of cheap terraced housing, comparable to much of the stock of the inner city, may still be found in peripheral boroughs as socially diverse as Bexley, Waltham Forest, and Richmond.

In contrast to outer London, the older housing stock of inner London means that owner-occupation is frequently associated with poor housing conditions and lack of amenities. This is particularly true of areas of terraced housing such as that found in Tower Hamlets and Newham (see Map 3.5). This form of housing stock is more closely associated with home ownership of manual workers—including significant numbers of coloured immigrants. In fact, within inner London, the rates of owner-occupation are higher for West Indian immigrants than they are for the British-born population of the same area. However, large-scale slum clearance in many parts of inner London has severely depleted the stock of poorer quality housing for sale, thereby putting home ownership further out of reach of low-income families. This gap has been accentuated by inflation and the property boom, which have been particularly acute in the south-east. Between 1966 and 1972 the average price paid for existing dwellings rose from £5000 to £12 000 in London and the south-east, while trends in the prices of new dwellings rose even more sharply.

Gentrification is a trend which is having a direct effect on the social geography of the inner city, as an increasing number of middle-class and professional households buy houses in the inner city for renovation and modernization, often with the aid of home improvement grants. These trends are not important in the traditional working-class areas of east London, but in selected districts such as Islington, Stockwell, and Fulham, many houses which are structurally solid have reverted to middle-class owner-occupation after some decades of subdivision and multiple-occupation in lodging-house accommodation for low-income groups (see Map 3.4).

3.2 Owner-occupied households

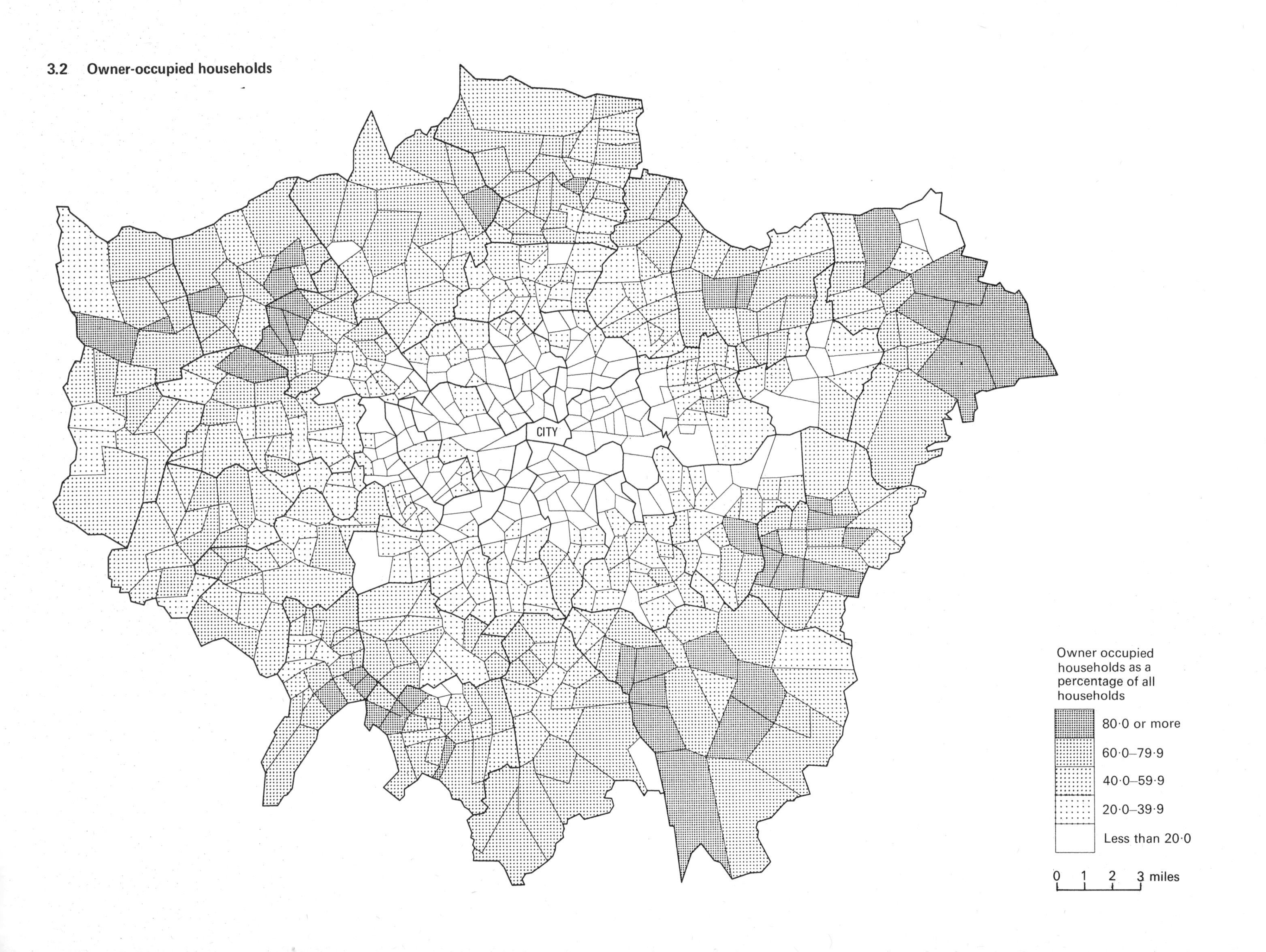

3.3 Council housing

Within Greater London one household in four rents its accommodation from a local authority. But, as Map 3.1 shows, the distribution of this tenure group differs sharply throughout the urban area. A major belt of local authority housing extends eastward from the southern parts of Islington and Hackney and the northern sections of Lambeth and Southwark to incorporate Tower Hamlets, much of Newham, and almost all of the borough of Barking. In fact 69 per cent of all households in Barking are in council accommodation—a marked contrast with Kensington and Chelsea, which falls at the other end of the spectrum with less than 8 per cent of its households in public housing.

To a large extent these spatial variations reflect differences in the housing policies of individual local authorities. Programmes for public housing are intrinsically related to political aims, social attitudes, and the extent to which policy-makers regard public housing as an essential welfare service. But, within inner London, there is also a close link between the proportion of council housing and the historical development of the urban area. In particular, much of post war activity in public housing has been concentrated in areas of acute housing stress, and large areas of nineteenth-century slum housing have been cleared in boroughs such as Tower Hamlets and Southwark. It does not necessarily follow that those who are displaced by such clearance schemes will be rehoused in the same area, or indeed in council accommodation at all (see 3.4), so that programmes of redevelopment and public housing may alter the social environment as well as the physical landscape. Increasing building costs, especially the rising price of land, has meant that there has been a significant decline in the number of new dwellings constructed since the mid-1960s. Nevertheless, the high proportion of purpose-built flats means that most council tenants are in dwellings which are structurally sound and have a full range of amenities, although high rise flats pose obvious problems for families. Exceptions to this generally high standard of housing are found in slum areas which have been purchased for compulsory clearance or redevelopment.

Local authorities have traditionally directed their housing programmes towards well-defined groups, especially large families in the lower income groups. Consequently, most heads of household in council housing are in manual occupations and include many skilled manual workers. Other needy groups, especially the old, have not been given high priority on the waiting lists, although many have been re-housed through clearance programmes. Housing policies are commonly backed by allocation procedures which give high 'points' to the poor, the badly housed, and households with children. The length of the waiting lists, which stood at over 208 000 for Greater London in December 1972, and the strict residential qualifications which many local authorities apply, discriminate against the young, students, and newcomers to the city such as coloured immigrants. One effect of these housing policies, together with the size of many housing estates, is that wards with high rates of council accommodation tend to have very distinctive social characteristics.

Council housing also has a direct effect on the residential mobility of large sections of the population. This is most apparent in the development of outlying housing estates, into which many families from inner London have been moved. In this sense council housing may be seen as the manual workers' avenue to suburbanization, although workers dependent upon the inner city for employment may prefer council accommodation in the central area. On the other hand, applicants on the waiting list for council accommodation may find that the residential qualifications incorporated into local authority allocation policies may discourage high residential mobility. In other words, such regulations may act as institutional barriers to population mobility. The effect of large public housing estates in outlying wards can be clearly seen in the map of local authority tenants, and, in a negative manner, in the patterns of owner occupation in the outer boroughs (Map 3.2). New Addington (south-east Croydon) and Hainault ward (north-east Redbridge) are prime examples of peripheral wards, in which almost all households are in local authority housing.

The effect of clearance programmes on the stock of relatively low-cost housing available for purchase has been noted in the section on owner-occupation. One factor which has partly offset this trend is the sale of council property to existing council tenants. For example, between 1968 and 1972, the Conservative-controlled G.L.C. sold almost 4000 of its dwellings in the three north-eastern boroughs of Barking, Redbridge, and Havering, a policy which the Labour Party has reversed since it won control of the G.L.C. in 1973

3.3 Local authority rented households

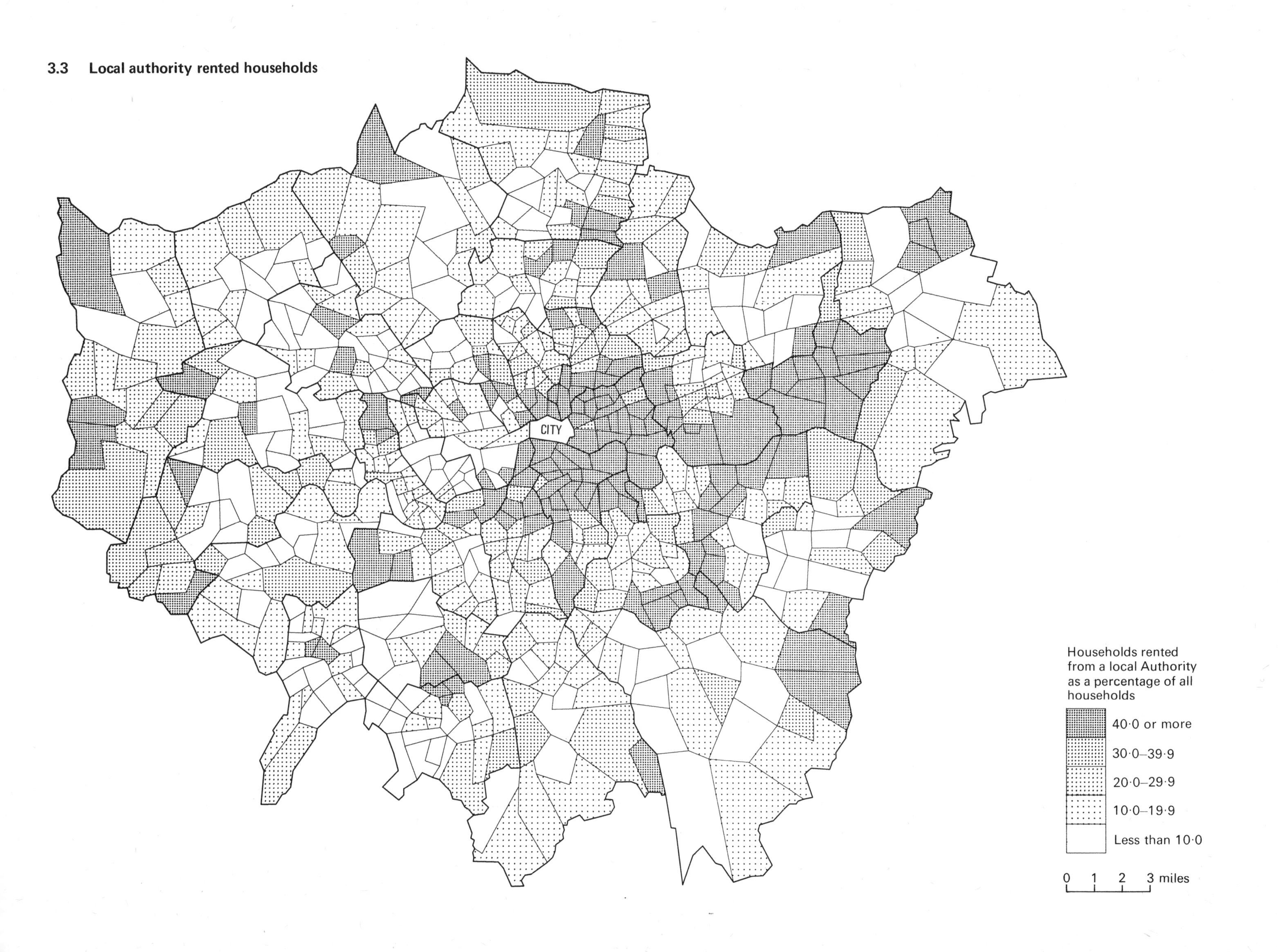

3.4 Private renting

The increases in owner-occupation and local authority housing over the past 60 years have been accompanied by a decline in privately rented accommodation. But one-third of London's households still depend upon the private landlord, and the vast majority of these households rent *unfurnished* rather than *furnished* accommodation.

The distinction between furnished and unfurnished accommodation has a number of important implications. The average weekly rent of furnished accommodation in 1971 was over £6 compared to £4 for unfurnished accommodation. But, although furnished accommodation is more expensive than any other form of tenure, the quality of housing in this sector is commonly very low. There are other disadvantages to furnished accommodation: less security of tenure; the ineligibility, under the housing regulations of many local authorities, of furnished tenants for re-housing from clearance schemes; and the exclusion of many of them from the 1972 Housing Finance Act provisions, which offer financial assistance to tenants who cannot afford a 'fair' rent. Although it is apparent that there is a ready market for accommodation in which the furniture is provided, many landlords also prefer to furnish their properties—however sparsely—because the financial return is greater and because it is easier to evict unwanted tenants.

In terms of the social geography of London, the furnished and unfurnished sectors are significantly different in their patterns of distribution, although both are concentrated in the inner city (Maps 3.4a and b). Furnished lettings are more concentrated to the north and west of the central area than are unfurnished tenancies, and no ward east of the City of London has more than 20 per cent of its households in furnished accommodation. With minor exceptions, such as the Balham-Streatham area, the highest concentrations of furnished lettings are found north of the Thames. On the other hand, households in unfurnished lettings are distributed more evenly throughout inner London. The major belt of concentration of this tenure group is also west of the City (from southern Camden to southern Hammersmith) but unfurnished tenancies form a high proportion of households in the east and to the south of the river, including areas such as Leyton in the north-east and Penge in the south-east. The housing stock associated with furnished lettings is commonly larger and structurally more solid than the smaller terrace housing, which is frequently associated with unfurnished rentals and, in part, the distribution of the two forms of privately rented accommodation parallels the distribution of these two forms of housing stock.

3.4(a) **Privately rented households: Unfurnished**

In addition to differences in their spatial distribution, furnished and unfurnished lettings tend to cater for different social groups and are characterized by differences in the provision of amenities and levels of overcrowding. Unfurnished tenancies have traditionally been associated with heads of households in skilled manual jobs, but this dependence has been declining in the face of the growth of subsidized council housing. On the other hand, furnished tenancies tend to cater for a range of diverse social groups, including young professional couples and single people, who are prepared to pay high rent for a flat or a bed-sit in a fashionable area close to the centre, as well as transients and many socially-disadvantaged people, such as the unemployed, the unmarried mother, newcomers, and the coloured immigrant, who are unable to gain access to the more favoured tenure sectors of the housing market. These groups are commonly forced to depend on overcrowded, relatively expensive, and often squalid lodging-house accommodation, in which the high level of physical stress can act as a powerful demoralizing force. For such accommodation, the rent is sometimes higher than that paid by a council tenant for a completely self-contained house or flat. Overall, the housing stock of the privately rented sector is older and in worse physical condition than the stock of any other tenure group. Its state is closely linked to the decline of privately rented accommodation, for in 1970 this sector accounted for 80 per cent of the expected demolitions planned at that time.

Private landlords, the suppliers of this tenure group, range from owner-occupiers, who let a basement or part of their house to help with mortgage repayments, to large property companies with more than 8000 lettings. An inquiry on housing in Greater London (the Milner Holland Report) showed that approximately 60 per cent of all private landlords had only one letting and that this group accounted for only 14 per cent of all tenancies. These landlords include many European and coloured immigrants letting part of an owner-occupied property. On the other hand, large companies (landlords with over 100 lettings) formed only 0·3 per cent of all landlords in London but controlled 32 per cent of all lettings. There has been concern over this sector because of an increasing number of speculative landlords encouraging multiple-occupation of dwellings without due concern for the welfare of tenants. This has opened the debate on the role of the private landlord in areas of housing stress. In this context, the Labour Party, which subsequently won control of the G.L.C., incorporated into their 1973 election manifesto proposals to municipalize some of the privately rented accommodation in London in an attempt to halt abuses and alleviate housing stress in the inner city.

3.4(b) Privately rented households: Furnished

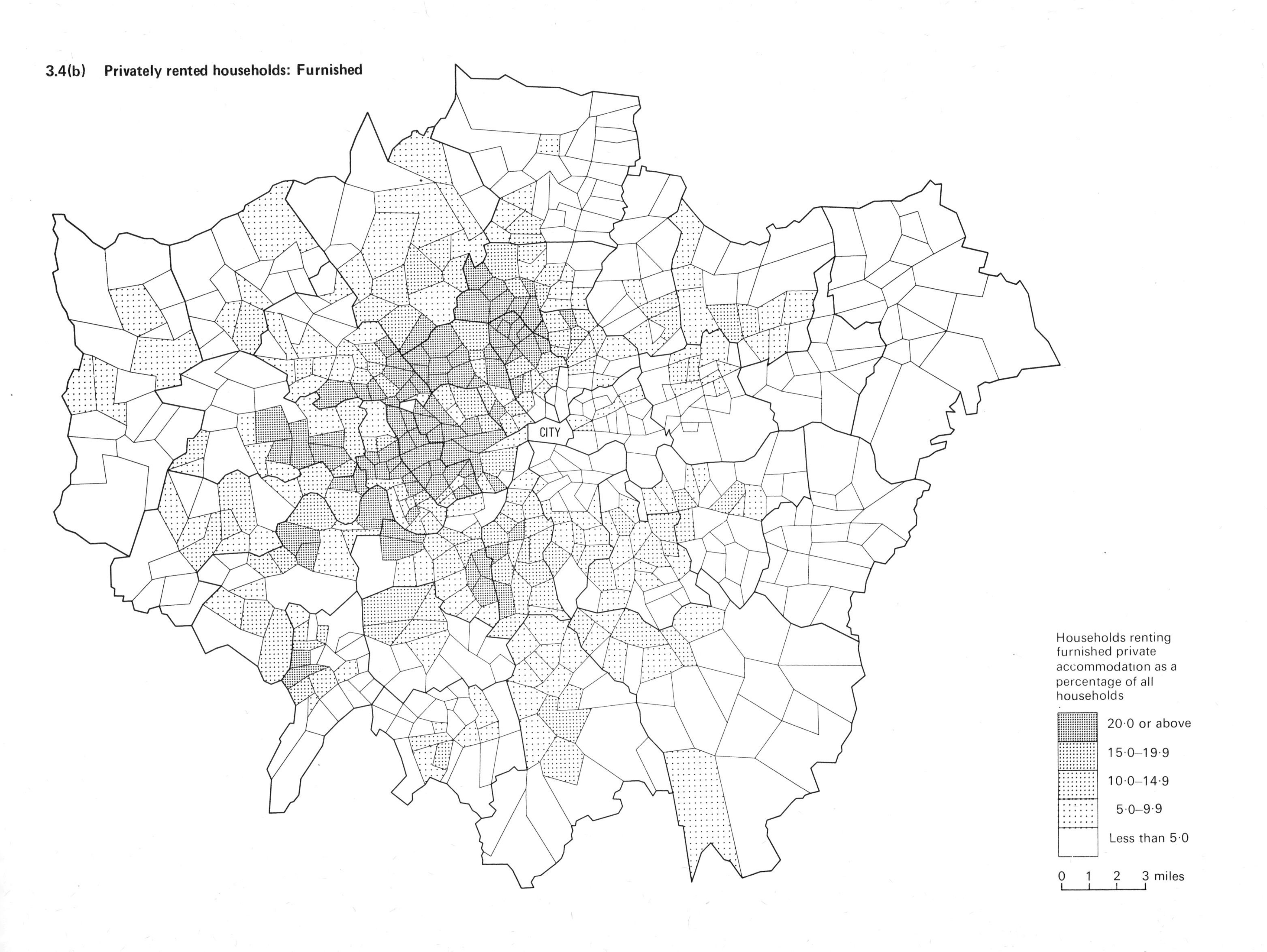

3.5 Housing conditions and overcrowding

There are three key housing factors which affect the quality of domestic life; the physical condition of the housing stock, the type of amenities available to the individual household, and the extent to which a household is living in overcrowded conditions. Some reference has been made to these aspects in the previous sections on housing in London, but it is appropriate to draw these strands together and examine them in their own right.

Houses which are classified as 'poor or unfit' are likely to have several defects, the most common of which are structural faults, dampness or leaking roofs, poor ventilation and lighting, and rotten woodwork. Most of the dwellings in poor physical condition also lack a bath and many have no inside lavatory. Although the physical condition of the housing stock is largely related to its age, the standard to which the housing stock was originally built varied considerably. For example, many small Victorian terrace houses or working-men's cottages were built cheaply to low standards, whereas the Victorian middle-class dwellings are structurally more solid, more spacious, and are more likely to be in fair condition today. The outcome of this has been that, for instance, in the working-class borough of Tower Hamlets, 71 per cent of dwellings built before 1875 are classified as 'poor or unfit' while only 7 per cent of the housing stock of a similar age in Westminster falls into this category.

Because of the associations between tenure groups and certain forms of housing stock it follows that the burden of London's housing stress is borne unequally between the different housing classes. Less than half of all privately rented dwellings are classified as being in 'good' condition (and 13 per cent are 'poor or unfit'), while only 1 per cent of owner-occupiers and 2 per cent of council tenants live in poor or unfit dwellings.

Basic domestic amenities and services, such as hot and cold water, and bathing and toilet facilities, are not found in all households or indeed, in all dwellings. Only one amenity variable, households *lacking an inside lavatory*, has been mapped here, but this distribution largely corresponds with areas lacking other amenities, such as baths, and with areas where there is a high proportion of 'poor and unfit' dwellings.

One London household in every twelve has no inside lavatory, but as the map shows, most of these are concentrated in the eastern and north-eastern sections of the metropolis in an arc from southern Waltham Forest to central and northern Southwark. Newham forms the core of this area and is the borough with the largest number of dwellings without either bath or inside lavatory. Other wards which score poorly on this

3.5(a) Households lacking an indoor lavatory

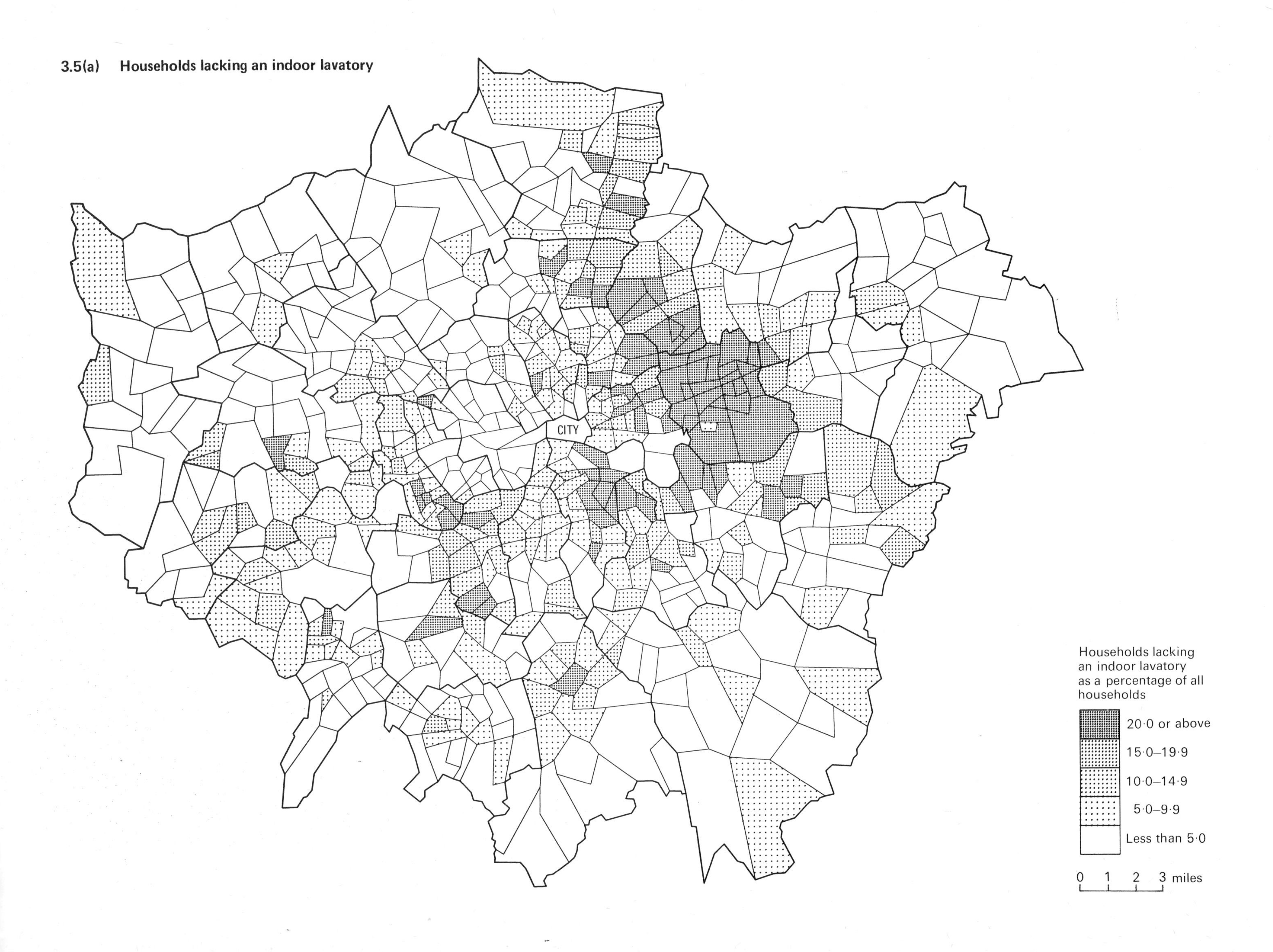

index are scattered throughout inner London (except in the higher social-class areas of Kensington, Westminster, and Camden), but some are found in peripheral boroughs, including Richmond, Croydon, and Bexley. Within all these areas it is mostly households renting unfurnished accommodation from a private landlord which lack the use of an inside lavatory.

Unlike poor amenities, *overcrowding* reflects a household's use of housing space, rather than a physical characteristic of the accommodation. A household is considered 'overcrowded' if there are more than an average of 1·5 people for every room which it occupies, but this is an arbitrary census definition, which takes no account of the size or condition of the rooms, nor the age and sex of the people inhabiting them. Nevertheless, there is a clear association between overcrowding and furnished lettings, especially bed-sit and lodging-house accommodation. Only 10 per cent of London's households are in privately rented furnished lettings, yet this housing sector contains almost half of London's overcrowded households.

The spatial association between overcrowding and tenure is clearly shown when the distribution of overcrowded households is compared with the map of privately rented furnished accommodation (Map 3.4). There is a marked concentration of overcrowding in inner London, especially in areas to the north and west of the central city. The reasons why overcrowding is so closely linked to furnished accommodation rather than to, say, unfurnished lettings, depend on a number of factors. These include the nature of the social groups which depend on this housing sector and the pressures on accommodation and land which stem from the housing shortage in London and from the paradox that many low-income and socially-disadvantaged groups have to find accommodation in those parts of the metropolis where land values are often very high.

Households with high rates of overcrowding include large families and groups such as unmarried mothers and coloured immigrants, who may have difficulties gaining access to favoured sectors of the housing market. But areas of overcrowding also coincide with the bed-sit areas such as South Kensington.

This section has highlighted some of the problems of London's housing conditions. Two final aspects should be noted, which provide contrasting pictures of the future conditions of London's housing stock. The first is that the useful life of much of the 'poor and unfit' housing could be extended through the provision of basic amenities such as baths. Home improvement grants provide some financial support for such provision but, in the privately rented sector, these improvements commonly push rents beyond the means of the low paid. The second factor is that, while the gap between existing households and available dwellings is slowly being bridged, increasing numbers of old houses are becoming 'unfit' as they reach the end of their useful lives. This ever-present tendency ensures that the problems of housing will remain a central social issue in London.

3.5(b) Overcrowded households

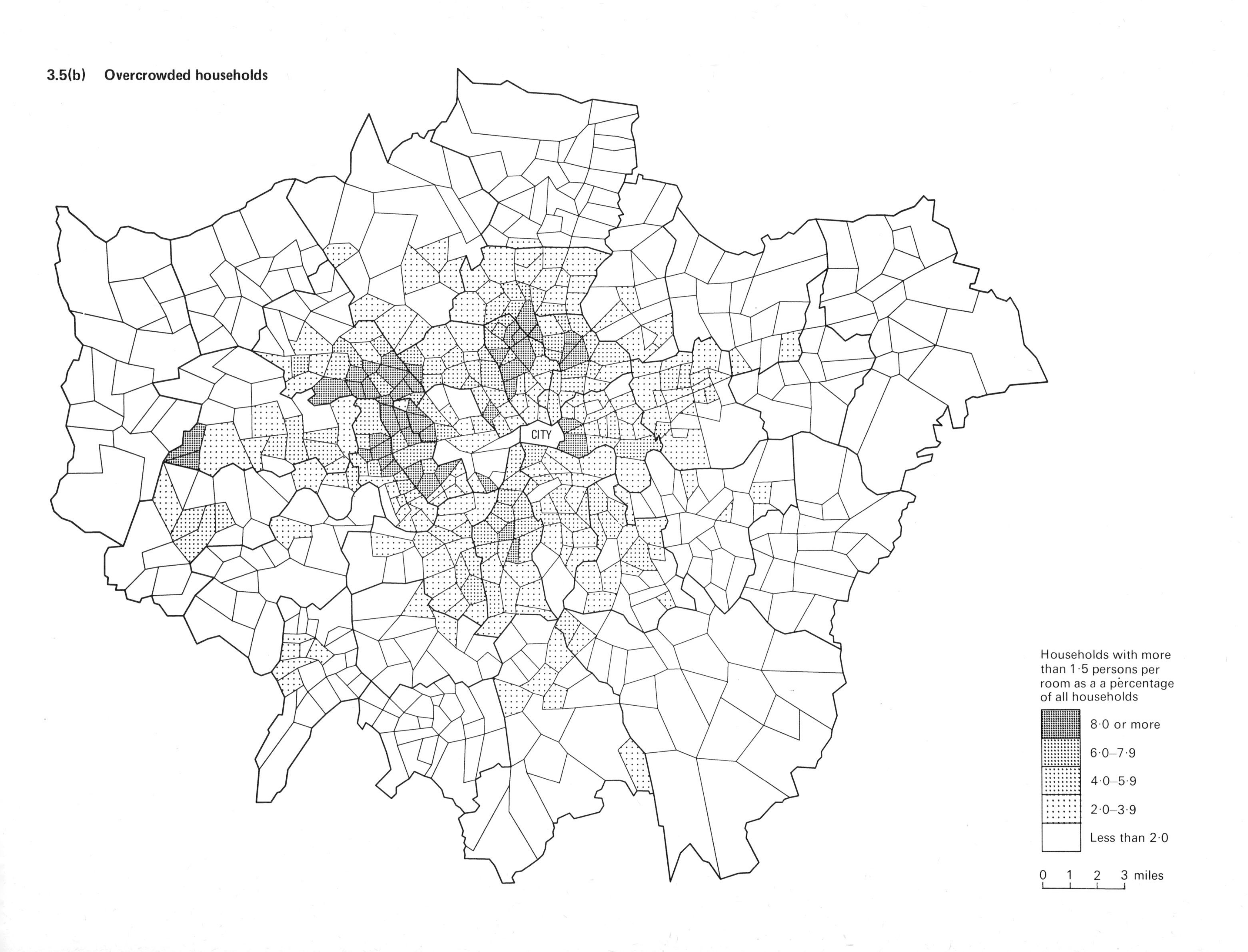

Population structure and immigrant groups

4.1 Age and sex structure

Throughout this century, the population of the south-east region has grown at a much faster rate than that of Britain as a whole, so that the region's share of the country's population steadily increased from 27·5 per cent in 1901 to 31·2 per cent in 1966. This movement of population to the south-east, frequently termed the 'drift to the south-east', seems, however, to have levelled off between 1966 and 1971.

Despite the rapidly growing population in the south-east region, London's population has been falling since the war, as large numbers of people have moved out to other parts of the region. Between 1951 and 1971, the population of Greater London fell by 10 per cent from 8·2 million to 7·4 million, while that of the rest of the South-East Economic Planning Region increased by no less than 39 per cent from 7·0 million to 9·8 million.

Not all London boroughs have shared equally in this population decline. Whereas many inner London boroughs lost more than 15 per cent of their populations between 1961 and 1971, some outer boroughs, notably Hillingdon, Bexley, Bromley, and Croydon, gained in population over the same period. This out-migration, particularly from inner London, is of two sorts: voluntary movement, and planned overspill resulting from the redevelopment of parts of inner London. Movements of a voluntary nature are selective, in that the younger and more affluent tend to move out, leaving the old and poor. The latter are forced to remain in inner London, either because of irregular work hours or because they cannot afford long-distance commuting. It is against this trend of out-migration, much of it selective, that borough variations in age and sex structure must be viewed.

The map opposite shows the age and sex structure of each borough by means of population pyramids. Each horizontal bar represents the proportion of the borough's population in five-year age groups, the area to the left of the vertical axis representing males, that to the right females.

The age variations between the boroughs, though generally small, are significant and provide an interesting background to the social characteristics of different parts of London. The most distinctive area of the city in terms of age structure is formed by the boroughs in the north and west of Inner London, namely, Islington, Camden, Westminster, and Kensington and Chelsea. These boroughs all have a preponderance of people in the 15-29 and, to a lesser extent, 30-44 age brackets. This is an indication that central London attracts young people and is matched by the number of flats and bedsitters in these boroughs. It is interesting that two of the four boroughs, Camden and Westminster, also have relatively more old people than London as a whole.

The 15-29 age group is also well represented in a number of boroughs surrounding this area. In two of them, Hammersmith and Wandsworth, the number of old people is again above the London average, while in the rest, Haringey, Lambeth, Brent, Ealing, and Hounslow, the population structure is more oriented towards young families, with above average numbers in the under-15 and 30-44 age brackets. As a result of post-war housing development, young families are also characteristic of many outer London boroughs, notably Hillingdon, Croydon, Bromley, Bexley, and Havering.

There are many middle-aged and old people in the other outer boroughs and in Greenwich and Tower Hamlets. But the social characteristics of these boroughs vary from the middle-class commuting areas of, for example, Richmond, Kingston, and Barnet in the west, to the predominantly working-class boroughs of Tower Hamlets, Greenwich, Barking, and Waltham Forest in the east. In the latter group, the under-15 age group also tends to be well represented.

The highest ratio of females to males is found in the south-west and in the western part of central London, reaching a peak in Kensington and Chelsea, where women outnumber men by a ratio of 55:45. Males form a higher proportion of the population in the sectors extending east and west out from the centre.

These geographical variations in age and sex structure in London have important implications for planning the provision of services, especially those for particular age groups as, for example, old peoples' homes and schools. The problem is whether facilities should serve the present population or be planned for future needs. Secondary schools sufficient for the existing school population in a borough may be grossly insufficient in the future. On the other hand, planning secondary schools on the basis of a high infant population may result in over-provision if many people leave the borough.

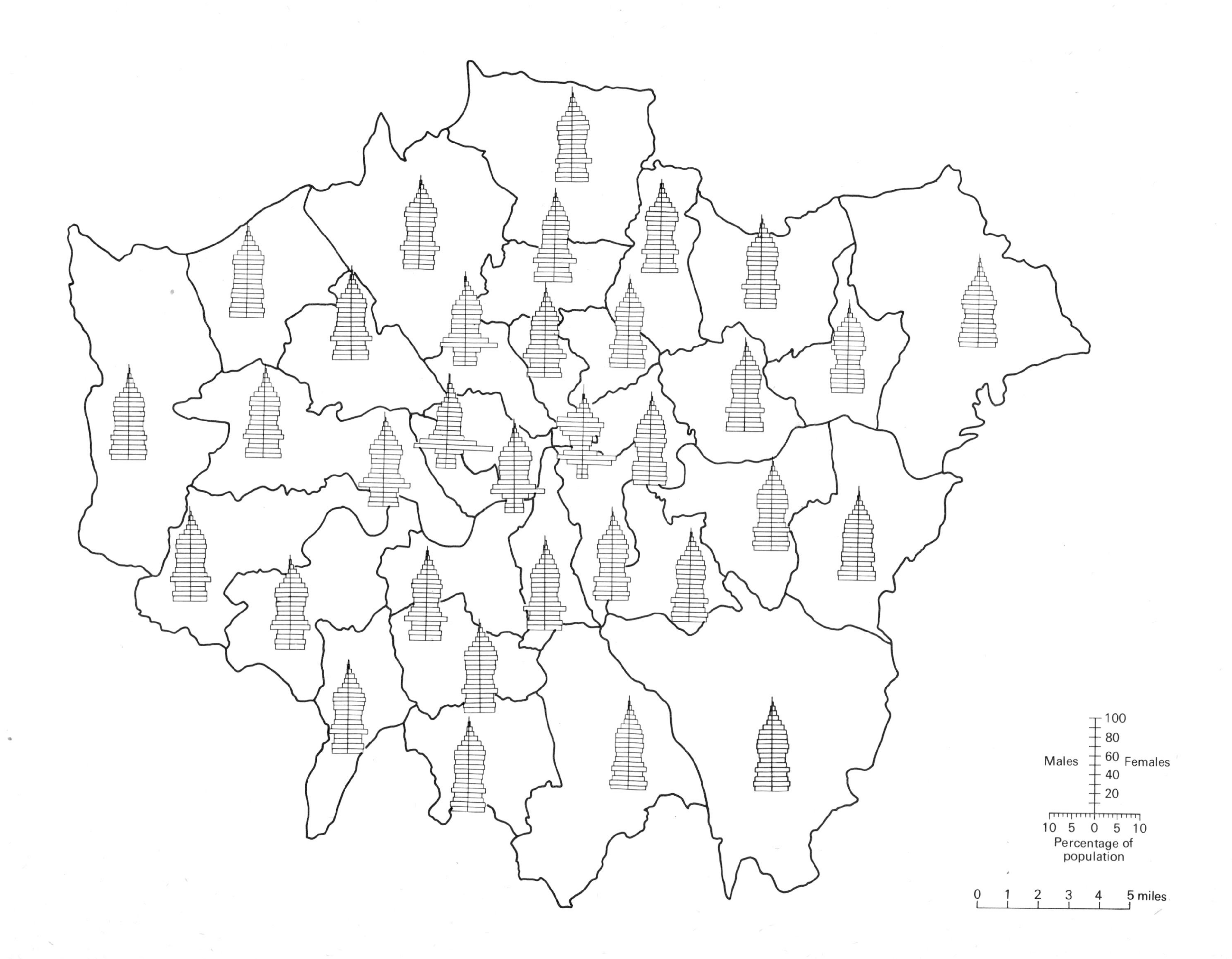

100
80
Males
60
Females
40
20
10 5 0 5 10
Percentage of population
0 1 2 3 4 5 miles

4.2 Residential patterns of immigrant groups

Greater London has been a traditional migration destination for many immigrant groups. These include the Irish, European Jews, and more recently, coloured immigrants from Asia, the Caribbean, and other parts of the Commonwealth. While this analysis is based on birthplace data, it should be noted that birthplace is not a direct surrogate for racial or ethnic status. There are, for example, many people resident in England who were born in India to British parents, and there are many coloured children born in Britain to parents from Asia, Africa, and the Caribbean.

All three immigrant groups considered here exhibit a high degree of residential concentration caused by several sets of factors. The classic explanation relates ethnic clustering to the immigrants' low economic standing in the labour market, and their inability to gain access to the most favoured sectors of the housing market.

It is only in the decaying residential areas of the inner city that the immigrant is supposedly able to overcome the two great problems facing the newcomer to the city—finding employment and accommodation. Immigrant concentrations commonly develop in the 'zone of transition', where former suburbs of high social standing slip into areas of multiple-deprivation. This can be seen most clearly in the distribution of the West Indian-born population in 1971. Map 4.2a shows that there are three main areas of concentration, approximately equi-distant from the city centre. Most outer boroughs have extremely low West Indian-born populations, as do the higher social-class areas of inner London in South Kensington, Westminster, and Camden. Much of the housing stock in the areas of highest concentration is obsolete, shared accommodation is common, and many households lack basic amenities.

While the presence of coloured immigrants in these areas is an effect of their decline rather than a cause, this cause-effect association between colour and bad housing is seen in reverse by many of the majority population. The coloured community of the inner city is commonly blamed for a physical environment, which has become obsolete, not through the settlement there of coloured immigrants, but through the passage of time. This reinforces latent feelings of hostility toward the newcomers; and these negative attitudes of hostility and discrimination may serve to reinforce the ethnic concentrations.

There is, however, a strong case for arguing that, in the British context, especially in London, it is misleading to suggest that racial prejudice and discrimination are the root causes of segregation. Rather, it should be

4.2(a) Residential pattern of West Indians

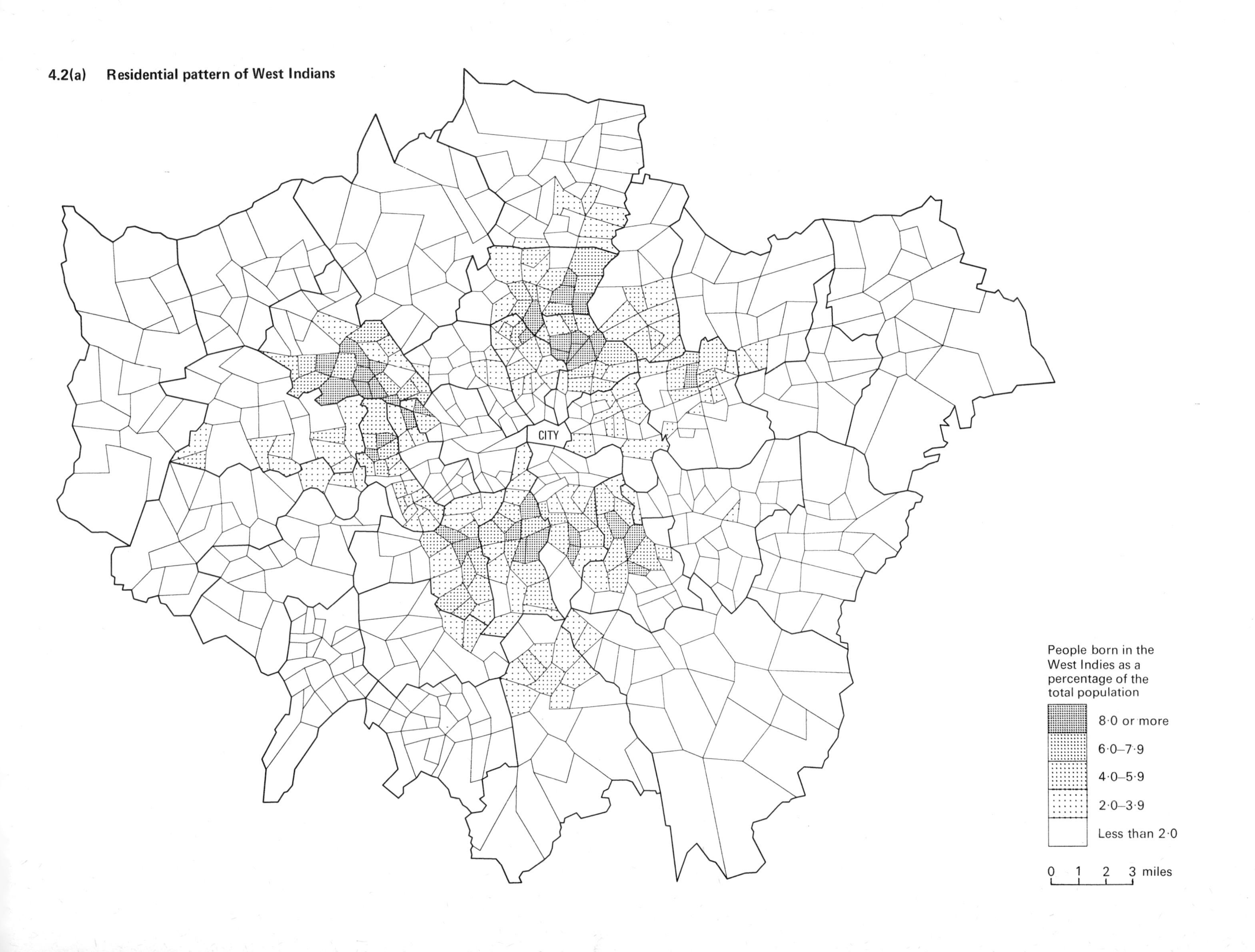

4.2(b) Residential pattern of Asians

4.2(c) Residential pattern of Irish

argued that housing shortages, general poverty, and periodic unemployment are the root causes of conflict within a broad social class which faces common problems and competes with little power or wealth for scarce resources. As the history of Irish and Jewish immigrants shows, any outside group which introduces further competition for scarce resources may generate conflict. But the emotive factor of racial differences between the coloured immigrants and the majority population acts as a focus for the manifestations of this conflict and they tend to assume a significance, which masks the underlying causes of the problem. The current distribution of Irish immigrants in London (Map 4.2c) shows a heavy concentration in the north-west (Brent) and west (Hammersmith) which suggests that residential clustering is caused by pressures faced by both white and coloured immigrant groups of similar low economic and social status. Other white immigrant groups in London (such as the Cypriots) also display relatively high levels of residential concentration and segregation and this reflects the fact that many immigrants have unskilled manual and service occupations and that there are very few middle-class immigrants. Among the British-born population different classes also tend to be residentially segregated. These social and economic factors are compounded by housing shortages in London, which promote the multiple-occupation of dwellings. Accommodation in such lodging houses is frequently let through ties of kinship and friendship (especially in the case of Asian immigrants) and this also tends to promote residential clustering.

There may also be positive forces within the ethnic community, which tend to encourage residential clustering. These stem from cultural and language differences between the host society and the immigrants' home country. This means that the spatial concentration of an ethnic group may reflect the development of an ethnic *community*, rather than a mere clustering of residences. Within these communities the development of a range of services has often helped to preserve the cultural identity of some groups. These services may range from petty trade to the professional services of medicine and law, and commonly include religious institutions, ethnic newspapers, theatres, and associations. In addition, the informal social contacts which stem from living near compatriots also provide a cohesive force which has an obvious effect on residential patterns. The Asian communities in London have developed perhaps the widest range of ethnic services and these ethnic ties, which include strong unifying religions, have been an important positive force in the development of the concentration of Asian groups in London. More than any of the other groups considered here the residential clustering of Asian immigrants is due to forces of self-segregation.

While the levels of residential concentration for the Asians are relatively high they are not located principally in the inner boroughs (Map 4.2b). The western boroughs of Hounslow and Ealing form the nucleus of the most important Asian concentration in London. The proximity of Heathrow Airport to this concentration provides an important outlet for unskilled manual and service work. Within inner London the principal Asian concentration is in the East End (Spitalfields), where the clothing industry has provided many jobs.

This examination of the distribution of immigrant groups in London shows them to be concentrated in distinct districts. Such patterns highlight the relationships between the distribution of different types of housing stock, the distribution of social-class groups, ethnic group cohesion, and the social relations between minority and majority populations. As such, the distribution of immigrant groups in London provides a clear example of how the spatial structure of London (the built environment and the distribution of population groups) is closely linked to the social structure and the social relations of its inhabitants. With Map 4.3 the discussion considers the changing patterns of immigrant distribution and whether these groups are becoming increasingly concentrated in the inner city.

4.3 Changes in the pattern of immigrant groups

At present immigrant groups in London are distributed in distinct clusters. In the intercensal period 1961–1971 the number of coloured immigrants in London more than doubled and the proportion of the G.L.C. population formed by this group rose from 2·3 per cent in 1961 to 5·5 per cent in 1971. The real interest expressed in these increases, however, stems not so much from the numbers *per se*, but from the widely held belief that the incipient ghettoes believed to exist in many parts of inner London are becoming more entrenched. Here we examine this view in the light of changes in the distribution of the West Indian-, Asian-, and Irish-born populations in London between 1961 and 1971. To put these trends in their proper perspective we have also mapped changes in the total population of London over this period.

Between 1961 and 1971 the totals of the West Indian-born population of Greater London increased by some 73 per cent, the Asian-born population by 1·0 per cent and the Irish-born by only 0·2 per cent. If the suggestions are true that the concentrations of coloured immigrants are becoming increasingly focussed in inner London, it could be expected that the inner boroughs would have a much greater increase in immigrant population over this ten-year period than did London as a whole.

In Map 4.3b the actual West Indian-born population of each borough in 1971 has been compared to that which might be expected if its 1961 population had increased by 73 per cent—the rate of growth for the West Indian-born population for London as a whole. Maps 4.3c and d illustrate comparable trends for Asian and Irish immigrants. It should be stressed that these maps do not illustrate distributions at any one point in time, but express changes over a ten-year period, although one aspect of the distribution is indicated by symbols for boroughs with immigrant concentrations greater than the G.L.C. average in 1961 and in 1971. The maps relate the immigrant population *observed* at the time of the 1971 census to the immigrant population that might have been *expected* at that date—had the increase in the immigrant population been the same in the borough as in London as a whole.

Map 4.3b shows that in 1961 the percentage of West Indian immigrants in all Inner London boroughs except Greenwich was higher than the corresponding percentage for Greater London. In 1961 West Indian immigrants were also concentrated in Brent and Haringey, boroughs which adjoin Inner London. Between 1961 and 1971 only in Lewisham and Wandsworth in Inner London and Brent and Haringey did the concentration of West Indian immigrants increase, and by 1971 Newham had become a borough of concentration with a West Indian population greater than in most Inner London boroughs. In Camden, Westminster, Kensington and Chelsea, and Tower Hamlets there was an *absolute* decline so great that by 1971 their proportion of West Indian immigrants were less than that of London as a whole. In the remaining Inner London boroughs the rate of increase was less than in Greater London, but they have remained areas of West Indian concentration.

The trends for the Asian-born population (Map c) do not show such a clear contrast between the inner and outer boroughs as do those of the West Indian population. Inner London, which had the highest concentrations of Asians in 1961, has experienced heavy losses over this period. The relative increases are focussed in two distinct areas. The first of these is in the western boroughs of Hounslow, Ealing, and Brent, where the degree of concentration of Asians has increased. The other area of relatively high gains is in the east, where Newham and Waltham Forest are now boroughs of high Asian concentration. The widespread relative losses in many other boroughs might suggest, however, that there has been very limited movement by the Asian population away from areas of high concentration.

Between 1961 and 1971 the Irish-born immigrant population declined in Inner London. These losses were heaviest in the arc of boroughs from Islington to Hammersmith, where in 1961, one-third of London's Irish population resided. Like the trends for the West Indian population, and in contrast to those of the Asians, the boroughs of relative increase are widely distributed. In 1971 no borough had an Irish population which was 50 per cent more than would be expected given its 1961 population and the intercensal trends for Irish-born immigrants in Greater London, while deviations of this magnitude were observed for the other two immigrant groups. The highest increases occurred in boroughs to the north and west, perhaps reflecting a degree of peripheral movement from those sectors of inner London with the highest Irish concentrations.

From an examination of the trends in the distribution of these three ethnic groups it is clear that the distributions are changing and that relative losses of immigrants have been widespread in the boroughs of inner London. What cannot be established from these maps, however, is whether the degree of concentration in London as a whole has declined since 1961 or whether higher concentrations now exist in different locations. This situation can be clarified by comparing the proportion of each group's total population in London in 'concentrated' boroughs (defined here as boroughs where the ethnic group formed a larger proportion of the borough population than the total ethnic group did of the total population of London in 1961 and 1971 (see table). These 'concentrated' borough changes are indicated by stars and triangles on Maps 4.3b, c, and d which show that the changes in the distribution of the Irish and the Asians represent a relocation of areas of concentrated settlement. Although the West Indians were the group most densely concentrated at both dates the intercensal trends for this group reflect a substantial decline in concentration as well as a redistribution of its population.

Proportion of immigrant population in 'concentrated' boroughs, 1961 and 1971

Date	*West Indian*	*Asian*	*Irish*
1961	86·4	60·3	57·1
1971	70·4	63·2	57·2

The loss of immigrants from the inner boroughs may merely reflect the general decline in the total population of the area (Map 4.3a). This analysis shows, however, that the immigrant groups in London are leaving the inner city at a much faster rate than the indigenous population. Fears of ghetto formation in the inner city do not, therefore, appear to be justified by the trends between 1961 and 1971. However, the concentration of Asians in Hounslow, for example, indicates that this does not necessarily reflect a breakdown of residential clustering and it is possible for relatively high levels of segregation to develop in suburban areas.

The main reason for the relative decline of these immigrant groups in inner London can be related to changing tenure patterns in those areas and the associated pressures of redevelopment and rising house prices. Many of the boroughs of relative increase have areas of housing stock which is cheaper than comparable stock in the inner city. The inner boroughs which have experienced the greatest losses of immigrants have been important areas of cheap furnished accommodation—the tenure category in which coloured immigrants are heavily concentrated. But there has been a considerable decline in cheap furnished accomodation in the inner city as property is redeveloped for upper-income groups (gentrification) or commercial purposes (especially hotel and office development). These trends have undoubtedly forced many immigrant households (as well as many of their British-born economic peers) to find similar accommodation in other areas. The maps suggest that much of this redistribution of immigrant populations has been channeled to the north-eastern and north-western parts of London and, in the case of Asians, to boroughs in the west.

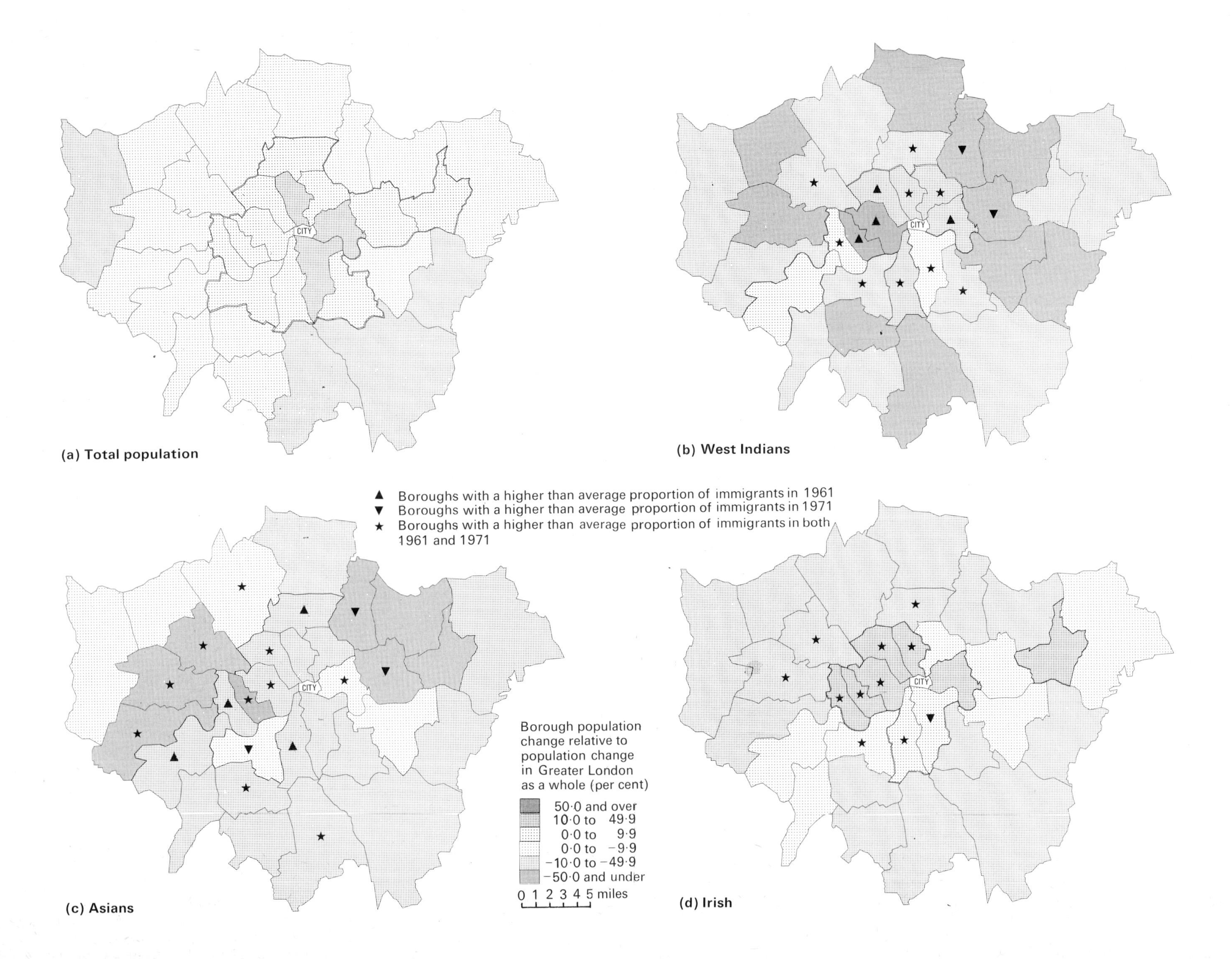
CITY
(a) Total population
(b) West Indians
(c) Asians
(d) Irish
▲ Boroughs with a higher than average proportion of immigrants in 1961
▼ Boroughs with a higher than average proportion of immigrants in 1971
★ Boroughs with a higher than average proportion of immigrants in both 1961 and 1971
Borough population change relative to population change in Greater London as a whole (per cent)
50·0 and over
10·0 to 49·9
0·0 to 9·9
0·0 to −9·9
−10·0 to −49·9
−50·0 and under
0 1 2 3 4 5 miles

Socio-economic groups

Social class is a difficult and controversial idea; indeed, it has been suggested that, with the general rise in living standards, differences in social class no longer exist. It is, however, evident that there are considerable differences between people in the amount of wealth they inherit, the income they derive from employment, the prestige attached to their jobs, and their ability to influence a wide range of political decisions. All of these attributes are embodied to some extent in the notion of social class but, because it would be difficult or impossible to reach an agreed and combined measure of all of them, it is necessary to use some single indicator to express the concept. The Registrar General in the census reports uses a person's *occupation* as an indicator of the social class to which he or she belongs.

Information on over twenty thousand occupations is reduced to just seventeen socio-economic groups (SEGs) on the basis of the status, income, and amount of skill attached to a job. These SEGs can be further grouped to produce a hierarchy of four occupational categories, broadly representing social class: professional and managerial, intermediate and junior non-manual, skilled manual, and semi-skilled and unskilled manual. The degree to which these four groups form distinctive geographical patterns, which are themselves related to patterns of housing, employment, retailing, and welfare services, is evidence of the continuing relevance of the idea of social class.

5.1 Professional and managerial workers

Professional and managerial occupations, the highest ranked occupational group, comprise employers and managers and professional workers whose occupations normally require a university degree or some other highly selective qualification. This group of occupations includes sales or personnel managers, architects, solicitors, and civil engineers.

Since we are using occupation as a measure of social class, those employed in professional and managerial occupations correspond to the upper middle and upper classes. People in these groups have incomes considerably above the average for all occupations and this affects their material well-being in a number of ways. Professional and managerial workers are more likely to own a house and a car, and their childrens' prospects are considerably brighter than most. For example, nearly 40 per cent of children of professional workers continue their education beyond the age of 19, compared with only 0·5 per cent of those of manual workers.

There are very wide variations in the geographical distribution of professional and managerial workers. Although the proportion of such workers for the whole of London is 14·4 per cent, the proportion for individual wards ranges from 46·5 per cent (in Cheam South ward, Sutton) to 1·4 per cent (in Cathedral ward, Southwark). In general, the wards with the highest shares of people in professional and managerial occupations are found in the outer boroughs. Even within these areas, however, there are large variations. In the south, a belt of wards with high proportions of professional and managerial workers extends from Richmond in the west, through Kingston, Merton, Sutton, and Croydon, to Bromley and south Bexley in the east. Another concentration of such wards is apparent in the north-west, covering the northern part of Hillingdon, Harrow, north Brent, Barnet, and west Enfield. There are also smaller clusters of wards with high proportions of professional and managerial workers in, for example, Wanstead and Woodford in the north-east, Havering in the east, Blackheath and Shooters Hill in the south-east, and Barnes, Putney and Ealing in the west. Many of London's professional and managerial workers actually live outside Greater London in places like Epsom, Guildford, and, even further afield in the home counties, although, in contrast, some choose to live nearer the centre of London, particularly in the desirable western parts of the central area. The whole of southern Kensington and Chelsea, as well as large areas of Westminster, like Hyde Park, Knightsbridge, and Maida Vale, are especially fashionable.

The pattern of residence of most of the upper middle and upper classes represents a preference for home ownership in a good, clean, spacious environment. Due to the high cost of land in the centre, such housing is only available in the outer parts of London, and even then only for those with higher than average incomes. However, as we have seen, many professional and managerial workers choose to remain in Central London. Many such people are single and are prepared to accept privately rented accomodation, albeit of a high standard, in order to gain the advantages of central London life. In fact, in recent years, the advantages of living close to the central area have drawn increasing numbers of the upper middle classes into inner London. Many old working-class areas have been 'invaded' and formerly shabby houses converted into elegant, expensive residences, often with the help of local government home improvement grants. Evidence of such conversion is provided by the general improvement in the decorative state of buildings, accompanying the change from private rental to owner-occupation.

In addition, the raising of the social status of an area is often accompanied by pressure from residents to improve the local environment. In Barnsbury, for example, this pressure resulted in measures to exclude through traffic from local streets. This process of 'gentrification' not only changes the social composition of many areas in inner London, but also results in visible changes in the quality of the local environment.

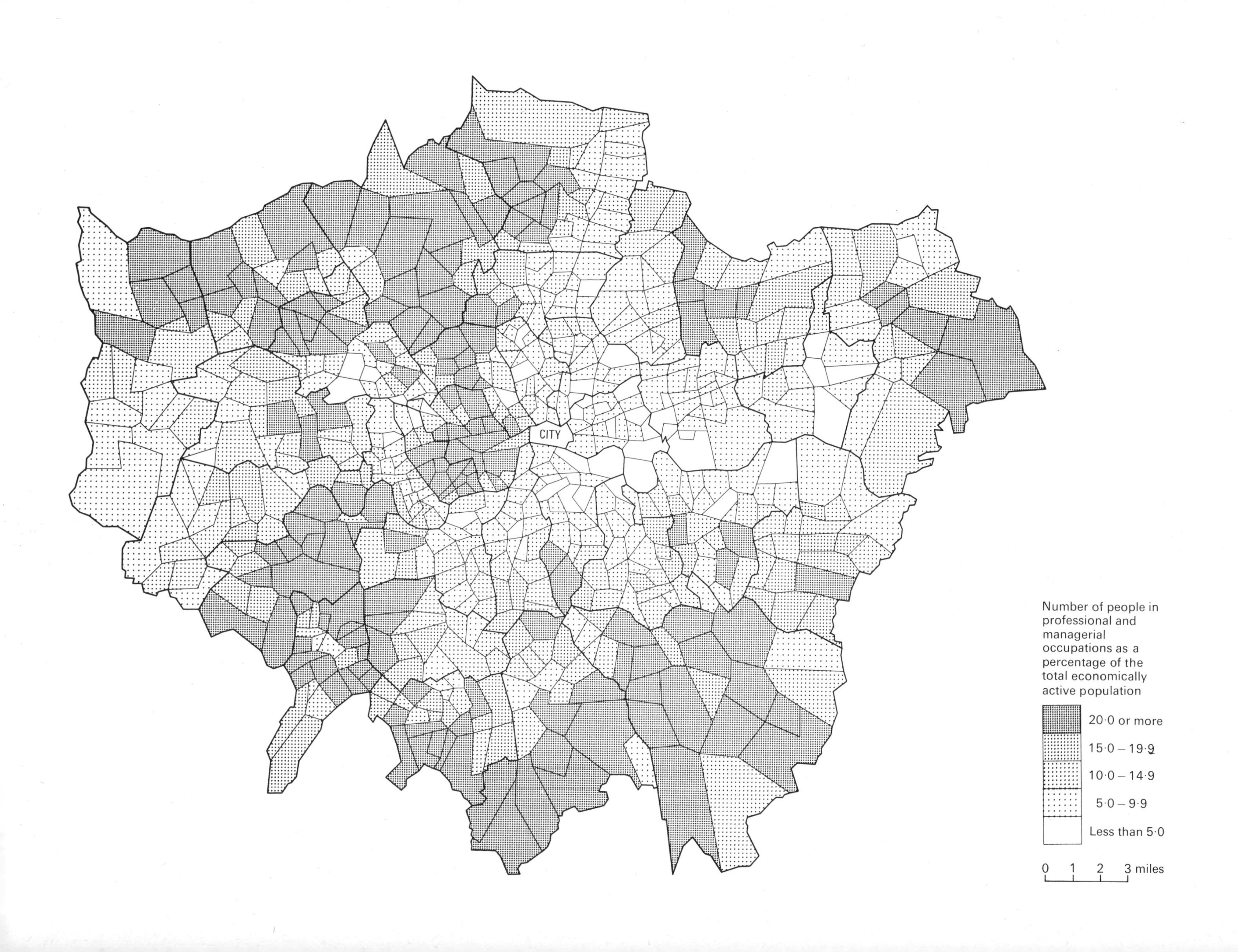
CITY
Number of people in professional and managerial occupations as a percentage of the total economically active population
20·0 or more
15·0 – 19·9
10·0 – 14·9
5·0 – 9·9
Less than 5·0
0 1 2 3 miles

5.2 Intermediate and junior non-manual workers

The second level of the occupational hierarchy comprises the two socio-economic groups of intermediate and junior non-manual workers. Intermediate non-manual occupations cover employees in non-manual occupations ancillary to the professions, and non-manual foremen and supervisors. Included in this group are teachers, nurses, social workers, and public health inspectors. Junior non-manual occupations comprise employees not exercising general planning or supervisory powers but engaged in clerical, sales and other non-manual occupations. This group includes clerks, office machine operators, typists, shorthand writers, and secretaries.

These groups, the lower middle classes, are less well-placed than professional and managerial employees in terms of income, housing, education, and welfare. In general, however, their incomes are higher than those of manual workers, although there is much overlap, with some manual jobs paying more than non-manual ones. Non-manual, or white-collar, jobs also tend to be more secure and offer better promotion and career opportunities as well as providing better sick pay, pension schemes, and working conditions; most non-manual workers enjoy a cleaner, less noisy, less dangerous, and generally more comfortable work environment. In the housing market, too, non-manual workers tend to be better placed since their greater job security increases their chances of obtaining a mortgage for house purchase although recent increases in house prices have now put owner-occupation beyond the reach of many non-manual workers.

Like professional and managerial workers, many intermediate and junior non-manual workers live in the outer boroughs. In the south, they are concentrated in much the same areas as the upper middle class namely Richmond, Kingston, Merton, Sutton, Croydon, Bromley, and Bexley. In the north-west, however, where professional and managerial workers are highly concentrated, there are fewer lower grade non-manual workers. Instead, the highest concentrations of the lower middle classes north of the Thames are found in the more modest north-eastern suburban areas of Redbridge and Havering.

Many of the lower middle classes also live in the older, inner suburban areas, at Crouch End, Fortis Green, and Muswell Hill in the north, for instance, Clapham and Streatham in the south, and Lee in the south-east. In addition, intermediate and junior non-manual workers are also concentrated in wards surrounding upper middle- and upper-class enclaves like Blackheath, Ealing, and Barnes.

Finally, the lower middle classes form a high proportion of the resident population of some wards in the west and north-west of central London. The highest concentration is found in the south of the borough of Kensington and Chelsea, but some pockets of lower middle-class residence are also found in Hammersmith and Camden.

As with the upper middle class, the movement out to the suburbs is the dominant trend in the residential pattern of the lower middle classes. As London expanded in the first half of this century, it was predominantly the middle classes, both upper and lower, who filled the expanding suburban areas. Whereas the north-west was the main area of upper-middle- and upper-class expansion, however, the lower middle classes moved out into the older, more modest, inner suburbs of south, north, and north-east London. More recently, the movement has been into houses built for owner-occupation in the newer suburban areas of Havering and Bexley in the east, and Bromley and south Croydon in the south, thus completing the pattern as we see it today.

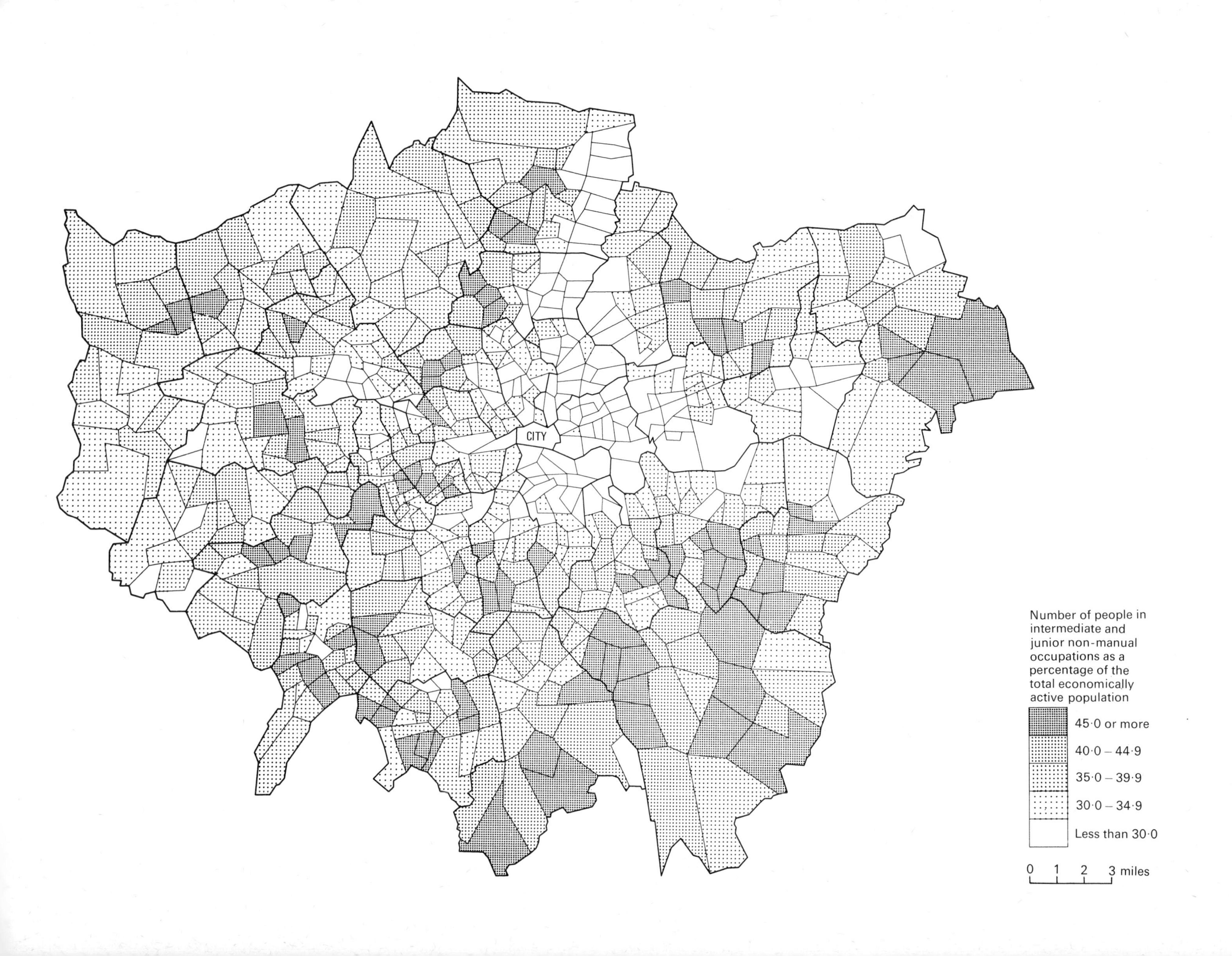
CITY
Number of people in intermediate and junior non-manual occupations as a percentage of the total economically active population
45·0 or more
40·0 – 44·9
35·0 – 39·9
30·0 – 34·9
Less than 30·0
0 1 2 3 miles

5.3 Skilled manual workers

Two socio-economic groups make up the skilled manual occupations: manual foremen and supervisors, and skilled manual workers, like bus drivers, machine tool setters, and electricians, who are engaged in manual occupations requiring considerable and specific skills.

In terms of income, the skilled manual workers, who comprise the upper levels of the working class, are, in general, less prosperous than white-collar workers, especially when fringe benefits, such as sick pay and pension schemes, are taken into account. The relatively low incomes of this group makes home ownership difficult, especially in periods of inflated house prices. As a result, most working-class people live in accomodation rented from either the local authority or a private landlord. Furthermore, much of the privately rented accomodation is of poor quality, often lacking basic facilities, such as baths and indoor lavatories. In education, too, the working classes are less well served than the middle classes. There is a sharp differentiation in educational performance between children whose fathers have non-manual occupations and those whose fathers have manual occupations. This is significant in that it is largely through education that the children of working-class fathers could obtain non-manual jobs, if they so desire.

The map of the distribution of skilled manual workers reveals two main concentrations, the larger of which is in the east and north-east of inner London, centred on Tower Hamlets and Hackney. This working-class area extends to the north up the Lea Valley, through Haringey and Waltham Forest to Enfield, and to the east through Newham and Barking to south Havering. The second concentration lies on the western fringe of the Greater London area, in the boroughs of Hillingdon, Hounslow, and Ealing. In addition, four other, smaller working-class areas can be identified: in south Brent and north Hammersmith, Southwark and Lewisham, and Kingston, while a discontinuous group of wards extends up the Wandle Valley from Wandsworth, through Merton into Sutton and north Croydon.

In general, the distribution of skilled manual workers corresponds to the distribution of manufacturing industry in Greater London. In the northern and eastern parts of inner London the rapid increase in the number of industrial establishments in the middle of the nineteenth century was accompanied by a great influx of workers, living in cheap housing built on and around the industrial sites. This was followed by a period of suburban expansion, which resulted in extensions of the working-class area eastwards to West Ham, Little Ilford, and East Ham, and northwards to Leyton and Walthamstow. The northward expansion has since been reinforced by industrial development in the Lea Valley dating from about 1900.

The working-class areas in the west are the result of twentieth-century industrial growth. The inter-war development of manufacturing industry at Park Royal, Wembley Park, and Harlesden accounts for the working-class areas of south Brent, while the largely post-war growth of industry along the main radial routes out of London and around London Airport has resulted in the growth of working-class residential areas in Ealing, Hillingdon, and Hounslow.

In south London, the working-class areas are similarly related to industrial development. Thus the growth of working-class areas in Southwark accompanied industrial growth in the nineteenth century, while those in the Wandle Valley and Kingston have been linked with the growth of manufacturing industry in these areas during this century.

A further element in the residential pattern of the working classes has been the location of local authority housing, particularly the old London County Council 'out-county' estates. These contributed towards the formation of the working-class areas of north, east, and west London, but some estates were located in parts of London in which there were few skilled manual workers. For example, the large working-class population in Addington, in the heart of middle-class suburbia on the Croydon-Bromley border, is the result of an L.C.C. 'out-county' estate. Such estates account for many of the 'anomalies' in the distribution of skilled manual workers in London.

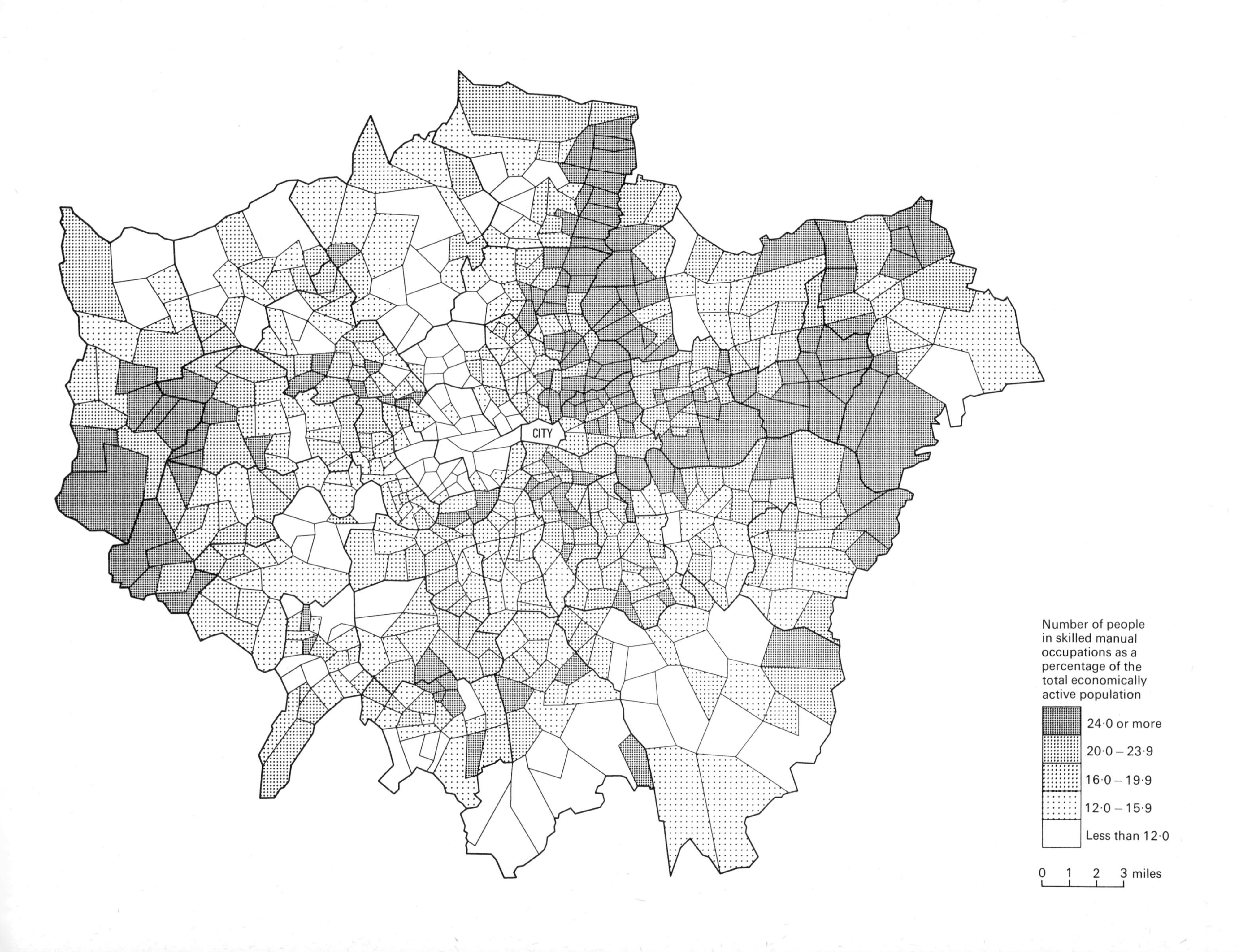
CITY
Number of people in skilled manual occupations as a percentage of the total economically active population
24·0 or more
20·0 – 23·9
16·0 – 19·9
12·0 – 15·9
Less than 12·0
0 1 2 3 miles

5.4 Semi-skilled and unskilled manual workers

The lowest level of the occupational hierarchy comprises the semi-skilled and unskilled manual workers, like bus conductors, machine tool setters, dock labourers, kitchen hands, and office cleaners. Most of these jobs are poorly paid and lack the security of higher status occupations, in part because of seasonal fluctuations in the demand for their labour. This makes the less-skilled manual workers' position in the housing market almost hopeless. As the repayments on a mortgage for house purchase would account for more than half of the monthly income of most semi-skilled and unskilled manual workers, they are almost wholly restricted to local authority or poor-quality rented accomodation. The children of less-skilled manual workers are also likely to suffer both in education and health. For example, the chances of such a child being unable to read by the age of seven are fifteen times greater than those of a child of a professional worker.

The residential distribution of semi-skilled and unskilled manual workers reveals a marked bias towards inner London, particularly the eastern part. South of the Thames, less-skilled manual workers are concentrated in an almost continuous belt of wards, often corresponding with those in which West Indian immigrants are concentrated, running from Wandsworth, through Lambeth, Southwark, Lewisham, and Greenwich, to the northern part of Bexley. To the north of the Thames, they are concentrated in the east, extending from Islington, northward through Waltham Forest and Haringey to Enfield, and eastward through Hackney, Tower Hamlets, and Newham, to Barking. The pattern is closely related to the development of manufacturing industries and the expansion of the Port of London. The rapid growth of these activities, in the late nineteenth and early twentieth centuries, was accompanied by a large influx of settlers. The nature of the employment meant that many of the immigrants were unskilled labourers and, owing to the irregular hours of work and the casual nature of much of the employment, most were forced to live in the immediate environs of their workplace.

Despite the post-war decline in the number of jobs in manufacturing and dock industries in these areas, the semi-skilled and unskilled workers remain because of the prohibitive costs of moving to areas of industrial growth. As a result, such workers have to seek employment in the lower paid service industries or become registered as unemployed.

Not all wards with high concentrations of less-skilled manual workers are, however, in the manufacturing and dock areas of eastern inner London. Many wards in Brent, for example, have high proportions of such workers, while in Camden, Westminster, Kensington and Chelsea, and Hammersmith, wards with more than average less-skilled manual workers are interspersed with wards in which the upper middle and upper classes are well represented. Indeed, in some cases, such wards actually coincide with those in which there are many professional and managerial employees. Such wards are said to exhibit social polarity. This situation, which is typified by Holborn ward in Camden, and South Stanley ward in Kensington and Chelsea, reflects the employment composition of central London. High-status office jobs in the City and the West End exist alongside the multitude of low-status service jobs, involving office cleaners, messengers, hotel kitchen staff, and waiters, all of which are necessary for the functioning of a commercial centre. In contrast, manufacturing industry and, more recently, routine office activities are moving out of the central area either to the suburbs or to other parts of the country. Although most of the high-status office staff, the professional and managerial employees, choose to live in the outer suburbs, some prefer to live in or near central London. On the other hand, the low status service employees, the semi-skilled and unskilled manual workers, have to live close to their work, because of their peculiar working hours and also because their wages are insufficient either to move out or to allow long-distance commuting. The middle income groups cannot, however, afford to live in the most desirable parts of inner London, and increasing numbers of them are refusing to tolerate the poor housing conditions in the less expensive areas. As a result, many such people move out to the suburbs, or even out of the London area, leaving concentrations of upper middle-class people, and upper and lower working-class people living in inner London, often in the same wards, but in vastly different social circumstances.

CITY
Number of people in semi-skilled and unskilled manual occupations as a percentage of the total economically active population
25·0 or more
20·0 – 24·9
15·0 – 19·9
10·0 – 14·9
Less than 10·0
0 1 2 3 miles

Jobs, shopping, and transport

6.1 Employment composition and change

London is by far the largest single centre of economic activity in Britain and its nearly five million workers constitute more than one fifth of the nation's work-force. But, like population, employment in London is declining; between 1961 and 1970, 300 000 jobs were lost, nearly 7 per cent of the total. This decline has been largely due to the decentralization of employment, both planned and unplanned, from Greater London to the Outer Metropolitan Area.

The decline in employment has not, however, been equal in all sectors of the economy. The greatest loss of jobs has been in manufacturing industries where, by 1970, the 1961 total had been reduced by 360 000, or more than 20 per cent. In contrast, employment in service industries has increased over the same period, due to the rapid growth in non-local services, i.e. those serving the south-east region or even the whole country. During the 1960s, employment in these services, namely insurance, banking and finance, professional and scientific services, and public administration, rose by nearly 20 per cent, an increase of 190 000 jobs. This growth has not, however, been shared by services fulfilling a mainly local function. In fact, employment in these services, namely gas, electricity and water, transport and communications, distributive trades, and miscellaneous services, declined by 2 per cent over the same period.

The effects of these trends on the availability of jobs in individual boroughs is partly a function of the relative importance of each economic sector in the borough. The first three maps opposite show the employment patterns of the main sectors of the economy in 1966, the latest date for which such information is available.

Jobs in manufacturing industries are concentrated in three areas (Map 6.1a). The most extensive covers north and east London, with the exception of the borough of Redbridge. The highest concentration is found in Barking, where 67 per cent of industrial employment is in manufacturing, of which nearly half is in the motor industry, chiefly at the Ford works at Dagenham. Enfield and Hackney, in the Lea Valley, also have high proportions of manufacturing employment, with engineering and electrical goods the leading sectors in the former and the older, nineteenth-century clothing and footwear and timber and furniture industries most important in the latter. The second main manufacturing area is in west London, centred on Brent and Ealing. Like Enfield, much of the employment is in the modern engineering and electrical goods industries, most of which are concentrated on the inter-war industrial estates at Wembley, Harlesden, and Park Royal. Such estates are also characteristic of the third manufacturing area, in south London, notably at Kingston, Merton, Mitcham, and Croydon, where engineering is also the predominant industry.

Employment in local services (Map 6.1b) is considerably higher in the inner areas of London but the nature of the employment differs greatly between the east and the west. In the east, concentrations of employment in docking and related activities account for the high ranking of Newham and Tower Hamlets. The high local services employment of the boroughs in the western part of inner London, on the other hand, comprises distributive trades and miscellaneous services, which include retailing and entertainment. Among the outer boroughs where local service employment is important, Hillingdon's position is due to the large employment complex at Heathrow Airport.

The distinction between inner and outer London is again apparent in the distribution of non-local services (Map 6.1c). These are concentrated in the western and central parts of inner London, notably in Camden and the City, although some boroughs in outer London—Bromley, Sutton, and Harrow—also have above average employment in non-local services, mainly in professional and scientific services.

Map 6.1d shows the changes in each borough's total employment, in relation to changes in London as a whole between 1951 and 1966. The most striking feature of the map is the decline in inner London and growth in outer London. In addition, it is clear that the heaviest losses have been experienced by those boroughs in the east and north east of inner London which are most dependent on the declining manufacturing and docking industries. The overriding problem facing these areas and, to a lesser extent, the rest of inner London, is the matching of the decentralization of population and employment. Although most inner boroughs are losing both population and employment, the employees who lose their jobs when a firm closes down or moves its factory into the suburbs are frequently those least able to move or commute to areas with good job opportunities. Many redundant manufacturing workers are consequently forced to join the already large number of unemployed in inner London or to obtain employment in the service sector and, in most cases, accept a reduced income.

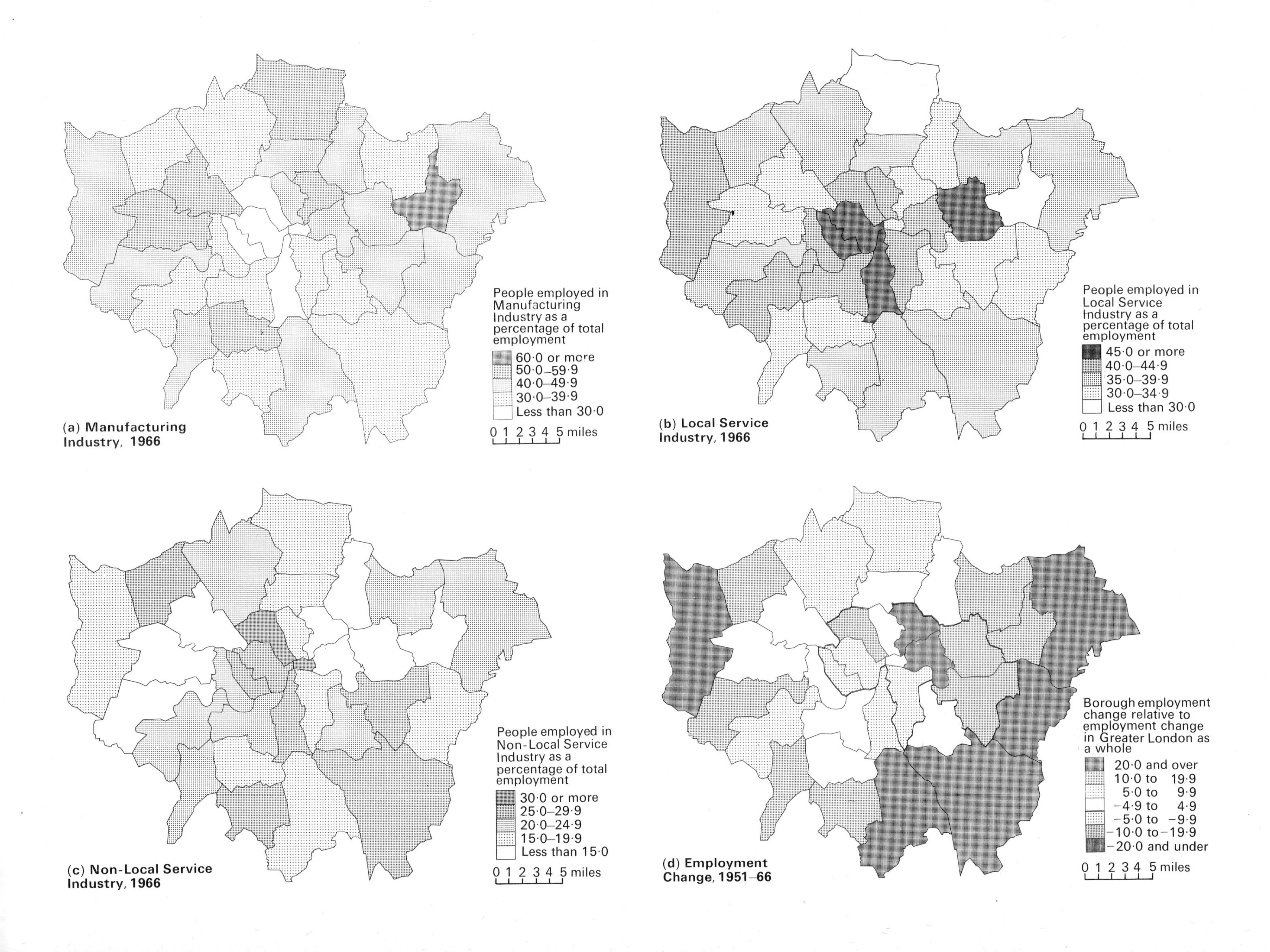
People employed in Manufacturing Industry as a percentage of total employment
60·0 or more
50·0–59·9
40·0–49·9
30·0–39·9
Less than 30·0
0 1 2 3 4 5 miles
(a) Manufacturing Industry, 1966
People employed in Local Service Industry as a percentage of total employment
45·0 or more
40·0–44·9
35·0–39·9
30·0–34·9
Less than 30·0
0 1 2 3 4 5 miles
(b) Local Service Industry, 1966
People employed in Non-Local Service Industry as a percentage of total employment
30·0 or more
25·0–29·9
20·0–24·9
15·0–19·9
Less than 15·0
0 1 2 3 4 5 miles
(c) Non-Local Service Industry, 1966
Borough employment change relative to employment change in Greater London as a whole
20·0 and over
10·0 to 19·9
5·0 to 9·9
−4·9 to 4·9
−5·0 to −9·9
−10·0 to −19·9
−20·0 and under
0 1 2 3 4 5 miles
(d) Employment Change, 1951–66

6.2 Shopping centres

The distinction between goods like food which are bought frequently and those like furniture which are bought only at long intervals gives rise to what is called a hierarchy of shopping centres. Such hierarchies have been observed both between towns and within them.

The map opposite shows the major shopping centres of London based on information gathered in 1966. The size of the circle representing each shopping centre is proportional to the area of floorspace devoted to shopping facilities, while the divisions within each circle express the relative importance of four types of shopping activity: department stores and supermarkets; food shops; non-food shops; and services and public houses. The map brings out the dominance of London's Central Area, which reflects its role as not only a regional but also a national shopping centre. Within the Central Area, broadly the area circumscribed by London's main-line railway stations, are a vast number of shops, which offer the consumer a wide variety of choice. Many of the shops are of a specialized nature and these are often located in well-defined 'quarters', as for example, fine arts and antiques in Bond Street and jewellery and silver in Hatton Garden.

Shopping centres outside central London can be divided into two groups. There are major centres, such as Croydon, Kingston, Ilford, and Romford, which not only have a greater amount of floorspace devoted to retailing, but also tend to have a higher proportion of department stores, chain stores, and supermarkets. In the other, smaller suburban centres, the amount of retail floorspace is lower and much of this is taken up by independent non-food stores and, to a lesser extent, food shops. The most striking features of the geographical distribution of suburban centres are first, the lack of large centres in north and north-west London, probably due to their high level of accessibility to the Central Area, and, second, the dearth of suburban centres of any kind in the eastern part of London, particularly in Tower Hamlets and Barking.

Recent years have witnessed two important and, in some ways, conflicting trends in the geographical distribution of retail facilities, namely decentralization and concentration. The provision of retail facilities is very susceptible to changes in the size and spending power of the catchment population. Consequently, as large numbers of inner London residents, particularly those with higher incomes, have moved out to suburban areas, shopping facilities in inner London have suffered a corresponding decline. The corollary of this has been the rapid rise in the turnover of suburban shopping centres, which have grown at an even faster rate than that of the Central Area.

Concentration has resulted from the general improvement in the standard of living of large sections of the population and from changes in the organisation of the retailing industry. Increased ownership of motor cars, refrigerators and, more recently, freezers, has allowed many people to make longer-distance, but more infrequent, shopping trips, thus encouraging the concentration of shops in the larger centres. At the same time, changes in the organisation of the retailing industry have seen the rapid growth of supermarkets, department stores, chain stores and, more recently, hypermarkets, at the expense of the small shopkeeper. The introduction of larger retailing units, deriving economies from the introduction of self-service techniques and from large-scale buying and selling, has severely reduced the competitiveness of the small shopkeeper, who, lacking such economies, has to charge higher prices for the same goods. Because of the higher catchment populations required by larger units, these changes have tended to lead to an increased concentration of such facilities in the major centres.

The parallel trends of decentralisation and concentration have profound implications for the provision of shopping facilities for lower-income groups in London, who are concentrated largely in the inner areas. Not only are such groups in an area of declining population and hence, declining shopping provision, but they are also unable to capitalize on the developments in retailing outlined above, because many of them do not have motor cars, refrigerators, or freezers; nor do they have the ability to pay cash for goods in discount warehouses. As a result, low income groups are forced to make frequent, short-distance trips to their local shops, where they have to pay higher prices than they would in the larger centres. Consequently, while large sections of the community have benefitted from changes in retailing, these have not included those groups most in need.

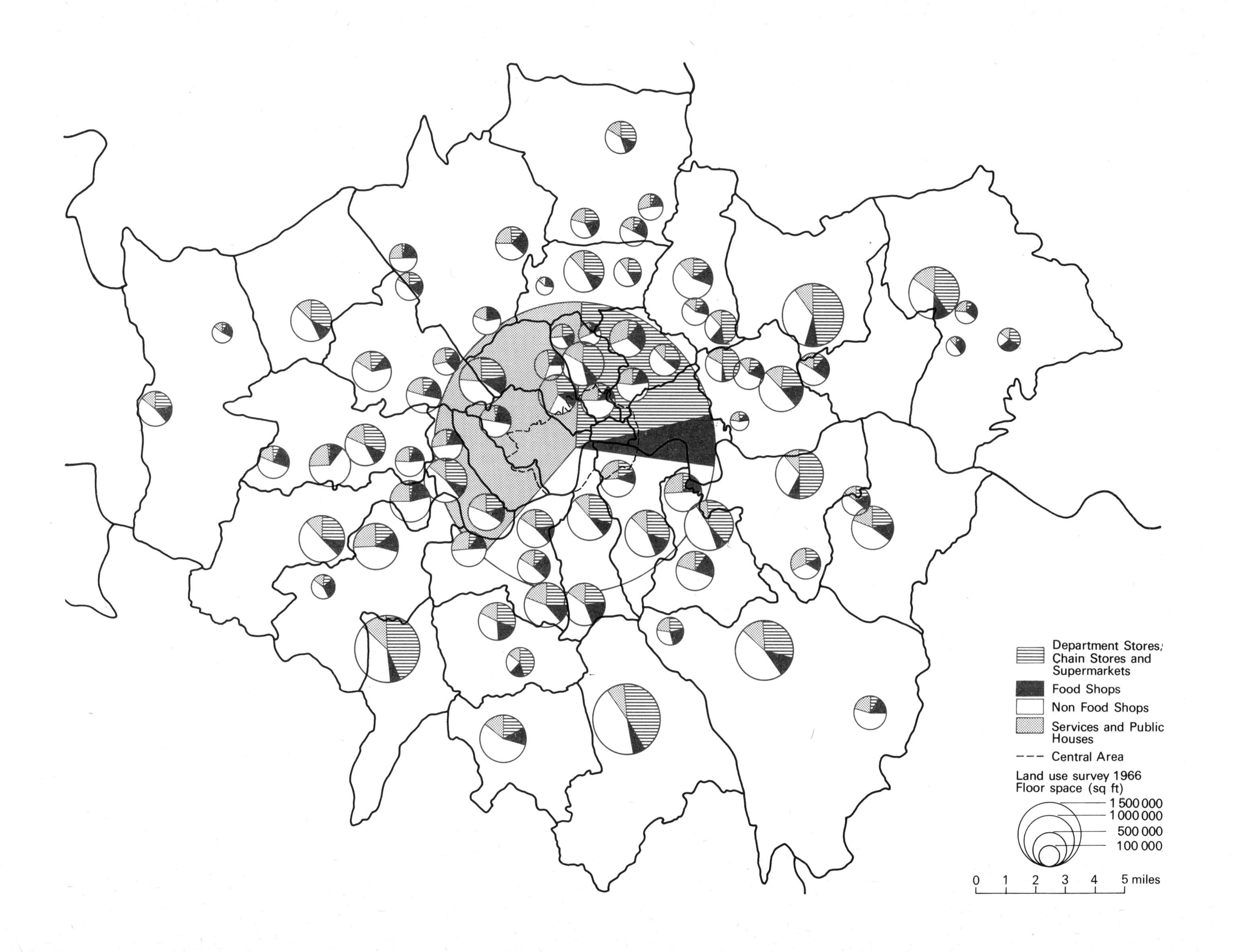

Department Stores, Chain Stores and Supermarkets
Food Shops
Non Food Shops
Services and Public Houses
Central Area
Land use survey 1966
Floor space (sq ft)
1 500 000
1 000 000
500 000
100 000
0
1
2
3
4
5 miles

6.3 Street markets

There is another side to shopping in London—the street markets. Many were established by Royal Charter in the City of London or in the outlying market towns. Others, particularly in inner London, were developed in the eighteenth and nineteenth centuries in a rather more haphazard way. The City of London had previously condoned trading in certain streets but, as the city expanded, costermongers first began moving their barrows from spot to spot and then settled in new and unauthorized locations. In due course these locations were recognized and were given a measure of sanction by London local authorities by the introduction of a licensing system for pitches at the end of the nineteenth century. The markets became strongly established in working-class areas, especially in the East End. Here a population, which included many Jewish immigrants and many who worked a six-day week, naturally favoured markets which operated on Sundays. The most famous of these are Club Row and Petticoat Lane.

The centres in the outer London boroughs which were once distinct municipalities have street markets with long histories. Bromley, Romford, and Enfield date from the thirteenth century, the latter receiving a charter from Henry III. Burnt Oak in Barnet, Leyton in Waltham Forest, Croydon, and Kingston were also granted royal charters. Most of these markets operate on one or two fixed days a week. This practice dates from the medieval fair and market, which was run by itinerant traders on a regular circuit. Today many of the traders in outer London still attend several markets on different days of the week. Despite the growth of shopping facilities in these outer centres, the markets still survive. The bulk of their trade is in fresh fruit and vegetables.

The distribution of streets where trading is allowed and the number of licensed stalls in each street are shown on the map. Most of them are concentrated in the traditional working-class districts of Camden Town, Islington, Hackney, Tower Hamlets, Southwark, Lewisham, and Newham. The western inner area has far fewer street markets and even these coincide with working-class enclaves in Pimlico and Church Street in Westminster, Portobello Road and Golborne Road in North Kensington, and Fulham in Hammersmith. Apart from the old-established markets, there is a striking absence of street markets in the outer boroughs, although, to a certain extent, private markets seem to fulfil a similar retail function. Private markets usually comprise a number of stalls in a courtyard adjacent to a main shopping street.

The economics of street markets explains their concentration in working-class areas. They attract both the casual trade and the merchant who cannot afford a shop—for instance, the stall holders who sell antiques on Saturdays in Portobello Road or Camden Passage in Islington. The cost of renting a pitch is low compared to a shop, although the best pitches are highly coveted. Some of the smaller markets sell only fruit and vegetables but the larger ones sell a whole variety of goods. For example, Romford market, which opens every Wednesday, Thursday, and Friday, has 585 stalls of which 15 per cent sell food, 3 per cent confectionery, 20 per cent clothing and footwear, 33 per cent household goods, and 20 per cent other non-food goods. Many of the goods are shoddily made or slightly damaged and are therefore on sale at reduced prices, which appeal to the lower-income groups.

This ability to meet an economic need and to adapt to changing economic circumstances explains the markets' survival. Their bustling atmosphere also holds loyalties, so that, for instance, former residents of Bethnal Green return week-by-week to Petticoat Lane from suburban Essex.

The peak years of the street market were the 1920s and the total number of street markets has declined since then. For example, Harrow Market with 272 stalls closed in the 1950s as did Chelsea Market with 60 stalls. The need for hygiene in handling foodstuffs and the increasing volume of traffic in the streets have contributed to this decline. Nevertheless, in 1966 some 13 000 people were still working as street peddlars and hawkers in Greater London, although this figure should be compared with the 23 600 costers and street sellers recorded by Charles Booth as working in the County of London alone in 1891. Recently it is the smaller markets which have declined while the larger markets have been increasing in size and, as a whole, street markets still comprise about a half of all London's outlets for fresh fruit and vegetables.

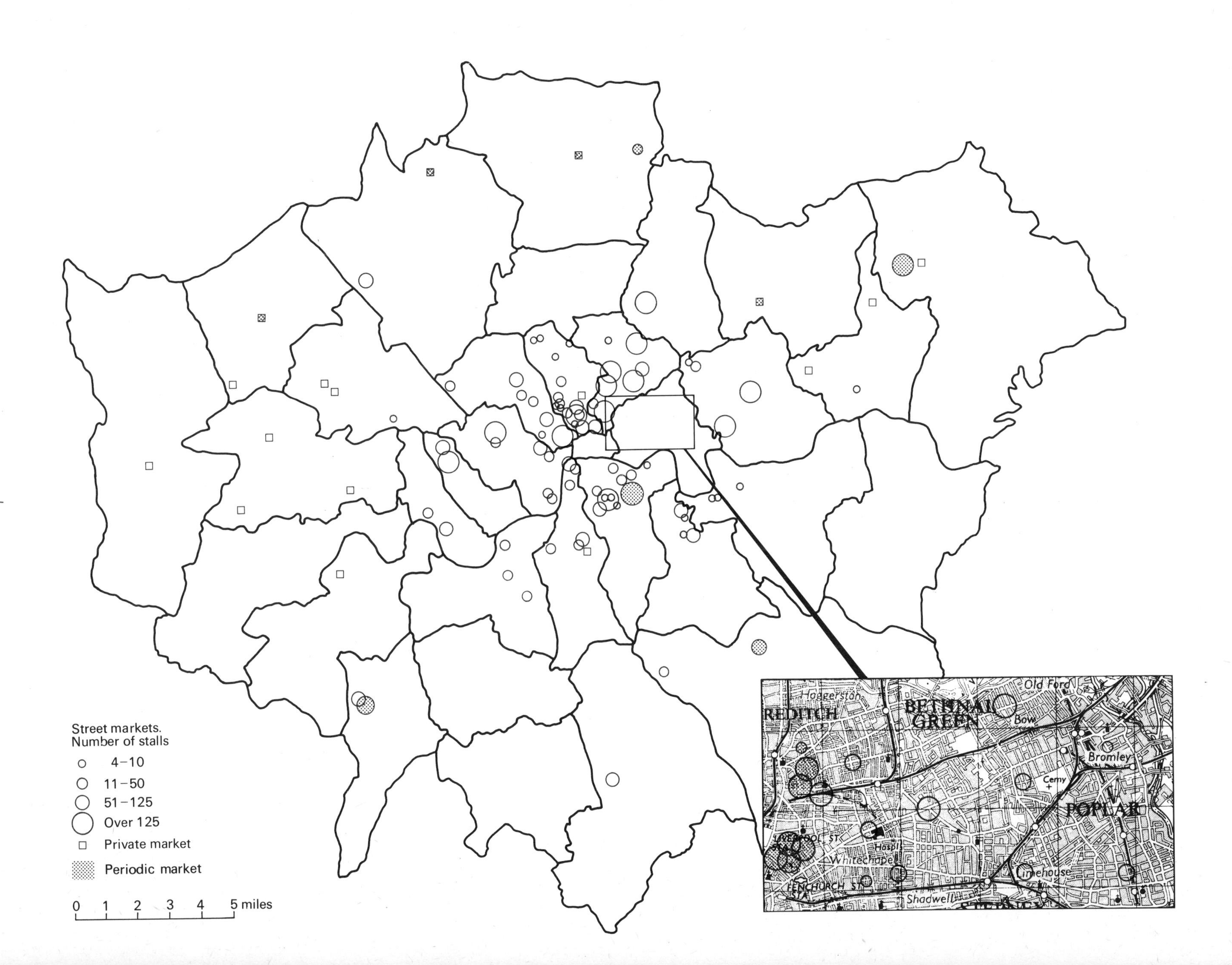
Hoggerston
REDITCH
BETHNAL GREEN
Old Ford
Bow
Bromley
Cemy
POPLAR
Hospl
Whitechapel
FENCHURCH ST STA
Limehouse
Shadwell
Street markets.
Number of stalls
4–10
11–50
51–125
Over 125
Private market
Periodic market
0 1 2 3 4 5 miles

6.4 Accessibility

So far in this section, we have established that there is some degree of geographical concentration in opportunities for employment and shopping. This is also true, in varying degrees, of the provision of educational, cultural, and health and welfare services. The accessibility of a ward, the ease with which its residents can travel throughout the urban area to find jobs and use these services, can either compensate for, or exacerbate poor local provision of jobs and services. This point has been made by protests in wards with poor access to jobs and services. In the Isle of Dogs (Millwall and Poplar ward), a series of such protests culminated in a 'Unilateral Declaration of Independence', in which the residents sealed off the area to draw attention to its poor service provision and inaccessibility.

Movement from one area of London to another depends on private transport by motor car or public transport by bus or railway (British Rail or London Transport underground). Map 6.4a shows the availability of private transport in each ward, measured by the number of households owning at least one car as a percentage of all households in the ward. The most striking feature of the pattern is the distinction between the inner and outer boroughs. In most of the wards in the outer ring of boroughs more than 50 per cent of households have at least one car, whereas in many inner London wards the figure is as low as 20-30 per cent. There are exceptions to this general pattern, such as Waltham Forest and Barking in outer London and Knightsbridge, Blackheath, and Dulwich (College ward) in inner London.

The difference in car ownership between inner and outer London matches the distribution of social classes to the extent that those outer London boroughs, like Waltham Forest and Barking, which are predominantly working-class, also have low car ownership. But social class does not explain everything. A recent survey showed that car ownership for any social class is higher in the outer suburbs. This can largely be attributed to first, the increased restrictions on car ownership in inner London, (e.g. residential parking permits), secondly, the higher level of public transport provision, and thirdly, the shorter journey lengths in inner London.

The measurement of public transport provision presents more problems. First, our unit of measurement is the ward and people cross ward boundaries to use public transport facilities in a neighbouring ward. Secondly, we are looking at quantitative, as opposed to qualitative, variations in the provision of public transport services. In other words, we

6.4(a) Car ownership

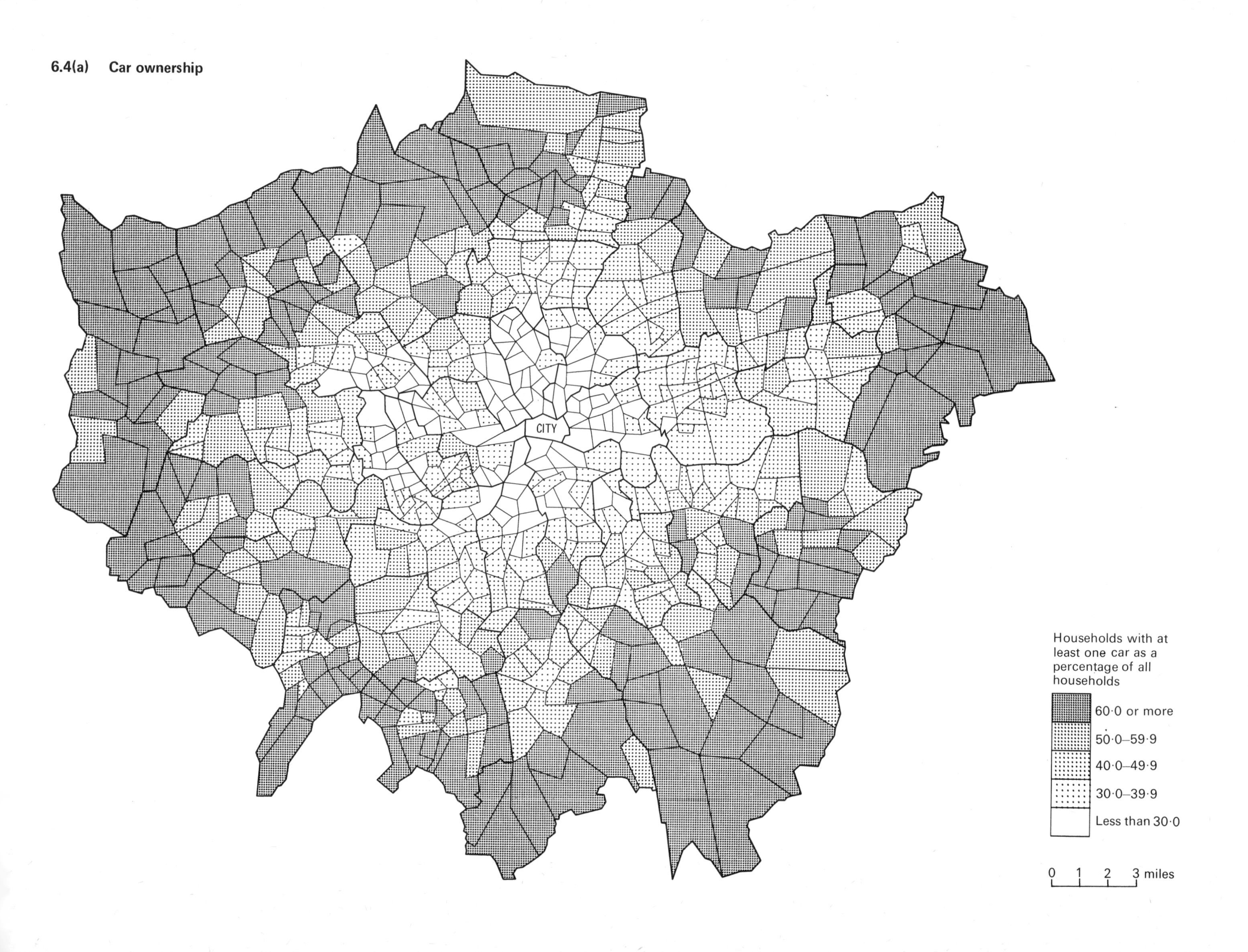

are only measuring how many bus services pass through, or along the boundary of each ward and the number of passenger railway stations (both British Rail and London Transport) within, or on the boundary of, each ward. In neither case is the quality of the service, e.g. overcrowding in the peak hours or irregularity in the off-peak periods, considered. Thirdly, it must be remembered that wards vary in area. Their boundaries are drawn in relation to population size and so the less densely populated outer boroughs are larger in area.

Bus accessibility is highest in the central area and declines towards the outer boroughs. There is, however, a marked directional bias in this pattern. Corridors of high accessibility extend from the centre towards Putney in the south-west, Wimbledon, Mitcham, and Croydon in the south, Lewisham in the south-east, Tottenham and Edmonton in the north, and Hammersmith in the west. These corridors of high accessibility reflect the largely radial nature of London bus services, many of which, in crossing from one side of the city to the other, pass through the central area. In addition, there are local peaks in accessibility corresponding to the major suburban centres, notably Croydon, Kingston, Hounslow, and Hampstead. Such minor peaks represent local bus services, which have recently been reinforced by the introduction of flat-fare suburban services which feed more traffic into local centres.

The outer suburbs are less well-provided with bus services, especially in the east, south-east, south-west, and north-west. Many wards in these areas have fewer than four bus services. Bearing in mind the large size of these wards, this demonstrates the skeletal nature of bus provision in outer London. However, large areas in the east of inner London, both north and south of the Thames, also have relatively poor bus accessibility, as well as the area flanking the Thames between the central area and Richmond and the area to the north-west of Hyde Park, extending as far as Wembley.

The pattern of rail accessibility, which shows many similarities to that of bus accessibility, directly reflects the historical development of the British Rail and London Transport underground systems. The development of radial rail services accompanied, and in many cases preceded, the movement into the suburbs of the middle classes. Its main function was, and still is, to carry suburban residents to their central London jobs. In fact, of the 1·1 million commuters arriving daily in central London during the morning peak hours, 75 per cent are carried by rail.

6.4(b) Access to bus services

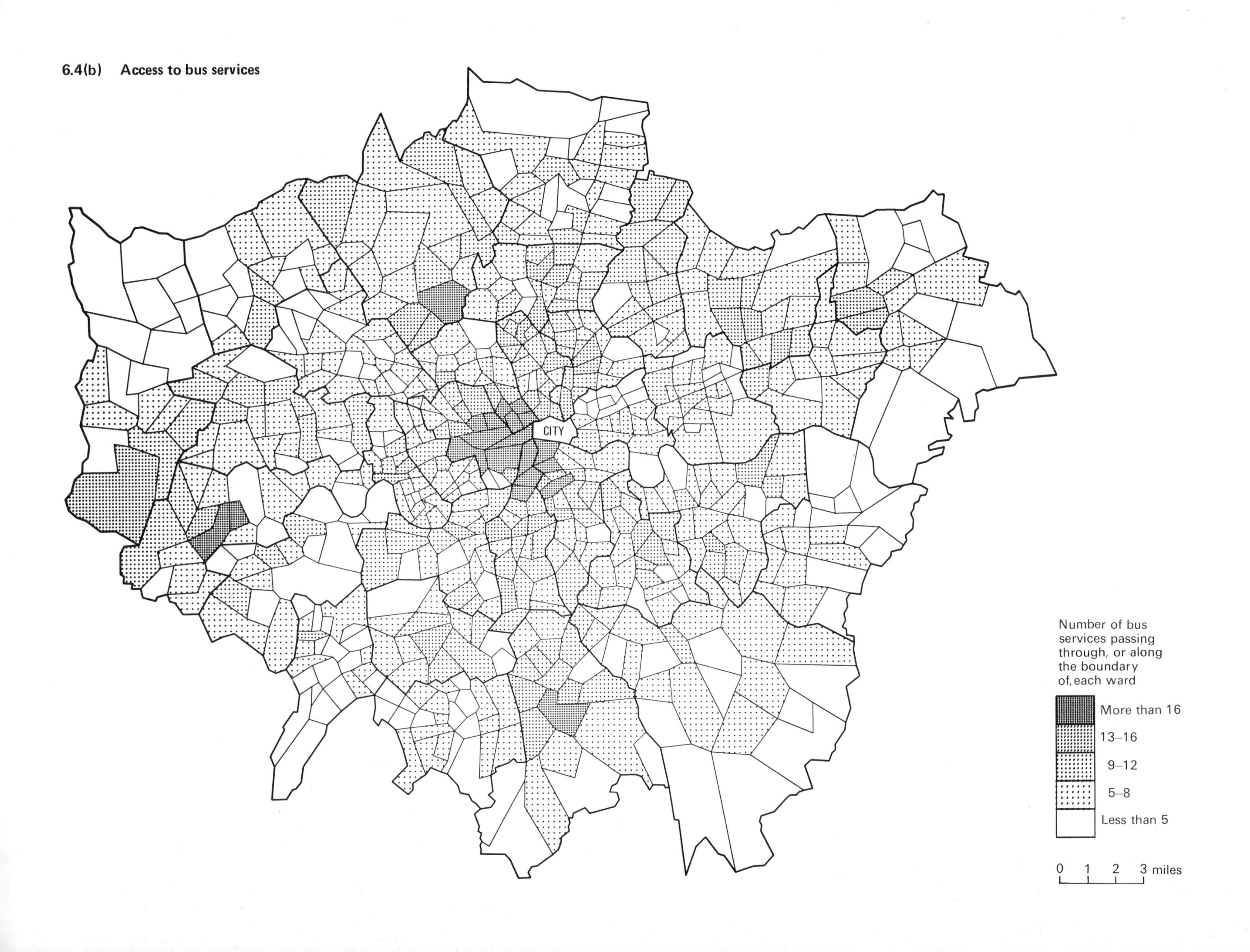

The outcome of this development is the pattern of rail accessibility shown on Map 6.4c:- a peak in the centre and sectors of high accessibility radiating out from the centre. Most striking among these are first, the broad sector extending to the west and north-west, based on the London Transport underground system, and secondly, the extensive area in the south and south-east, depending on the British Rail network. In addition to these are two other, more limited, sectors extending northwards to Barnet and north-eastwards through Leyton and Wanstead. This leaves large areas of London with comparatively poor access to the rail system, especially large parts of the boroughs of Tower Hamlets, Newham, Barking, Havering, Greenwich, and Bexley in the east, and the boroughs of Kingston, Richmond, Hounslow, and the southern part of Hillingdon in the south-west. In addition, large areas of north and north-east London, particularly in the boroughs of Enfield, Haringey, and Hackney, are without good rail accessibility.

Many of the wards which are poorly served by both forms of public transport have high car ownership. However, even where car ownership is highest, there are many households with no car, which depend on public transport. Furthermore, there will always be some sections of the community only able to use buses and trains—the young, the old, the infirm, and those who are unable to drive. Nor will all members of car-owning households have access to the family car. If the car is required by one member of the household, the other members are deprived of its use. Where the car is used for the journey to work, the household is effectively without a car throughout the working day. Unfortunately, because of the low population density and the declining potential market for public transport (as car ownership increases), the provision of public transport facilities for those who continue to rely on them will become correspondingly more difficult.

There are also many wards in inner London which have low rates of car ownership *and* are poorly served by public transport. They are concentrated in three areas: in the east, both north and south of the Thames; the north-west, around Harlesden, Willesden, and Kensal Rise; and in the west, in Fulham and Chelsea. Most residents in these areas find movement difficult and consequently must depend to a large extent on local jobs and services. These districts are often among the most socially and economically deprived in London. In the working-class area of east London, for instance, much of the housing is of poor quality, many of the schools are old and decrepit, retail provision is poor, and jobs in manufacturing industries and the docks are declining. Inaccessibility, therefore, adds to the multiple deprivation of these areas.

We have assumed that good public transport compensates for low car ownership, but this is not the case. Cars provide many advantages which public transport, at least in its present form, cannot match. A car owner has few restrictions on the timing of his journey, the route he takes, and the choice of his destination. Compare this with the public transport user, who is severly restricted by the services and schedules provided by the public transport authority. Whereas the car owner's mobility extends to all parts of London, families without a car can only travel with ease to areas with which they are linked by public transport; the radial nature of the bus and rail systems tends to restrict them to the sector of London in which they live. As a result families without a car have a narrower range of leisure activities, especially sport, fewer social contacts, and probably also less choice of job than car-owning families.

Even in areas in inner London with good public transport households without a car have only limited mobility. In view of the prevailing trends of decentralization of employment, retailing and, indeed, population, the provision of a higher level of mobility for those households which are forced to remain in the inner city consititutes one of the major planning problems facing London.

6.4(c) Access to rail services

6.5 Road accidents involving children

In the last section we saw the patterns of accessibility produced by the various transport systems operating in London—private motor transport, rail, and bus. There is, however, another aspect to transport systems, namely their side-effects, or social costs. Although noise and fumes are the most obvious, road accidents are the most serious.

The map on the opposite page shows the number of accidents on unclassified roads in each ward, involving injury to pedestrians aged less than fifteen. Using this map, we can look at the way in which the number of 'child accidents' varies between different parts of London. Accidents involving pedestrians on unclassified roads were chosen because they most clearly show the conflict between traffic and people since unclassified roads are designed largely for access within residential areas and consequently may be regarded as those on which people should be safest from traffic. Children, especially younger children, are more vulnerable since they are less aware of the dangers of road traffic. This accounts for the fact that of the 16 800 accidents in London in 1971 resulting in injury to pedestrians, 40 per cent involved children under the age of fifteen.

The distribution of 'child accidents' shows a clear distinction between inner and outer London; wards in inner London record considerably more accidents of this type than those in outer London. Within these two broad areas there are, however, considerable variations. Within inner London there are two main clusters of high accident wards. In the north and east one cluster is centred on Islington and Hackney, extending into Haringey, the southern part of Waltham Forest and, to a lesser extent, Tower Hamlets and Newham. In the south of the inner London area, another cluster runs from Wandsworth, through Lambeth and Southwark, to Lewisham. In contrast, wards in the west and north-west of inner London record considerably fewer 'child accidents', although small clusters of wards with bad records are apparent in the north of the boroughs of Hammersmith, Kensington and Chelsea, and Westminster.

In outer London, where, ironically, car ownership is highest, the situation is much better, although there are two accident clusters centred on Barking and Havering in the east, and Hillingdon and Ealing in the west. In addition, certain isolated wards in outer London have bad accident records, many of them corresponding with London County Council 'out-county' estates like Addington in Croydon.

We can explain this pattern by a number of factors relating to an area's traffic and environmental characteristics. The mileage of unclassified roads and the volume of traffic are directly related to the number of potential accident situations. Although the highest flows of traffic are normally confined to the main roads through an area, in busy parts of London many drivers use unclassified roads as short-cuts. The use of such routes, often termed 'rat-runs', can lead to large volumes of traffic on roads not designed for such purposes. The higher volume of traffic in inner London and the consequent use of 'rat-runs' accounts in part for the many 'child accidents' there.

The environmental characteristics of a ward are also important. The number of children in an area is clearly related to the frequency of accident situations. The smaller number of children in the west and north-west of inner London almost certainly explains the low incidence of child accidents in these areas. In addition, the provision of open space and play facilities affects the number of 'child accidents' in an area. A recent survey by the Metropolitan Police found that no less than 36 per cent of children involved in accidents were playing in the street at the time. The lack of open space in the northern, eastern, and southern parts of inner London is undoubtedly related to the number of 'child accidents'. In the outer boroughs of London, on the other hand, not only is there more open space, but most households also have their own garden. As a result, children in most suburban areas can play in safety, whereas in most parts of inner London, where traffic flows are higher, the street also has to serve as a playground.

CITY
Number of accidents involving injury to pedestrians aged less than 15 years on unclassified roads
More than 8
7–8
5–6
3–5
Less than 3
0 1 2 3 miles

Education, environment, and welfare

7.1 Dependent population

In every community there is a section of the population that can be described as dependent. The largest dependent groups are young children (under 15) and old people (over 65); the former have no income at all and the latter have fixed incomes. The young and the old require more welfare provision than most other sections of the community. This is not to say that everyone under the age of fifteen or over the age of sixty-five is dependent upon other members of the community. Some old people, for instance, have an income sufficient to guarantee their independence. By the same token, many other groups in society are not independent as, for example, unmarried mothers or fathers, the disabled, and the unemployed. In general, however, the young and old are the main dependent groups.

The map opposite shows the proportion of each ward's population under the age of 15 or over the age of 65. We are considering the two extreme sections of the 'population pyramids' which were examined in Section 4. For London as a whole the proportion of the population which we have called dependent is 34·4 per cent. It is, however, clear from the map that this figure varies considerably between different parts of the city. In some wards it is more than 42 per cent, while in others the figure is as low as 16 per cent. In general, a higher proportion of the population in east and north-east London is dependent than in the west. There are, however, some significant exceptions to this general pattern. Large areas on the outskirts of the eastern half of London, notably in Redbridge and Bromley, have low dependent populations, while in the west there are clusters of wards in which a high proportion of the population is dependent. The most extensive of these is in the south, centred on Sutton, Merton, Kingston, and western Croydon. Smaller groupings are also evident in Hounslow, Ealing, and the western part of Harrow. In addition, isolated outer London wards have high dependent populations. Many of these, for example Addington and Carshalton St Helier in the south, Yiewsley in the west, and Burnt Oak in the north-west, constitute old London County Council 'out-county' estates.

This pattern is the result of many, sometimes contradictory factors which affect population migration. Thus, for example, kinship ties found in the traditional working class areas of east London may be a constraint on the movement of families away from the area. On the other hand, the limited economic and social opportunities in such areas may encourage the out-migration of younger working members of the community. Variations in the availability of different types of housing, as they affect people in various stages of the life cycle, are also important.

Whatever the reasons, the geographical variations in the size of the dependent population have serious repercussions on the need for provision of welfare facilities of various kinds by local authorities. A large number of children in an area will require more schools, clinics, nurseries, and playgrounds, while the higher the number of old people, the greater the demand will be for homes and care for the aged. Furthermore, a high dependent population will increase the pressure on facilities provided for the whole population, notably hospitals and other medical facilities. This is because the young and the old tend to require medical attention most frequently. These groups make no financial contribution to the community, since the old have already contributed during their working lives and the young have yet to pay any tax. So, while the community is obliged to provide for its dependent population, geographical variations in the size of this group put severe pressure on those local authority areas in which the dependent population is most numerous.

7.1 Persons under 15 and over 65 years of age

7.2 Unemployment

Unemployment is a problem not commonly associated with London. In no year between 1961 and 1971 was London's unemployment rate more than three-quarters of the national average and in most years it was nearer one-half. Furthermore, London has always had a higher rate of job vacancies, in most years more than 20 per cent above that for Britain as a whole. Yet these statistics fail to tell the whole story. First, percentages conceal the fact that the *absolute* number of unemployed workers in London (196,000 in 1971) is comparable to that in Development Areas like Scotland and northern England. Secondly, geographical variations *within* London are such that some parts of the inner area have rates of unemployment as high as those in the Development Areas. Before looking at these geographical variations, it should be noted that the number of unemployed workers (i.e. those without work during the week of the census) recorded by the Registrar General is considerably higher than that given by the Department of Employment. This is due to the large number of people, especially women, who do not register as unemployed even though they are without work.

The map of unemployment rates in London wards reveals a strong contrast between the inner and outer boroughs. The central area of London is surrounded by an almost continuous ring of wards with unemployment rates of more than 6 per cent, while in most wards in outer London the figure is below 4 per cent. The pattern of variation in unemployment is clearly related to the distribution of socio-economic groups. Wards with high unemployment correspond closely with those in which manual workers, particularly those in semi-skilled and unskilled jobs, are concentrated. This relationship can be attributed in part to the effects of technological change involving the replacement of many unskilled jobs by machines and in part to the changing locational patterns of employment (see Section 6.1), which have resulted in a decline in certain manual jobs in inner London. Factory work, in particular, has declined as manufacturing firms have either closed down through mergers or rationalization schemes, or moved their operations out of inner London. For example, the closure of the AEI factory at Woolwich in 1968, following the merger of GEC and AEI, resulted in 5,500 people losing their jobs. Most of the displaced workers were unwilling to move away from the locality and many of them experienced a long period of unemployment.

This situation is aggravated by the Government's regional policy, which is designed to encourage manufacturers to move to regions of Britain with persistently high rates of unemployment. The Government has been very reluctant to grant permission for new manufacturing activities to develop in Greater London, even in localities of high unemployment. Employment in distributive trades, another traditional mainstay of London's manual workers, is also declining rapidly in the inner area, due partly to the movement downstream of the Port of London and its related activities. The decline in manual work, now a well-established trend both in London and the nation as a whole, stands in marked contrast to the sustained growth in administrative and, more especially, professional and managerial employment.

The spatial distribution of the unemployed in London can also be related to patterns of immigrant settlement, particularly those of the West Indians and the Irish (see Map 4.2). As several studies have shown, young male West Indians in particular experience disadvantages in the competition for jobs, due either to their lack of specific skills or to disguised racial discrimination.

The unemployed can, in some respects, be considered as part of the dependent population in the city (Map 7.1) in that they and their dependents receive financial assistance from the State in the form of various unemployment benefits. It must be remembered, however, that unemployment benefits are paid for out of the national *insurance* scheme, to which workers contribute. In addition, supplementary benefits are payable to the unemployed whose national insurance benefits have ceased or are not adequate for their circumstances. A *wage stop* rule operates in respect of benefits to ensure that an unemployed person does not receive more from the State than he would from full-time employment in his normal occupation. In London the minimum guaranteed wage is now (March 1974) calculated at £25.30 per week compared to £23.00 outside London. The differential of £2.30 per week reflects the higher cost of living in London.

7.2 Unemployment, April 1971

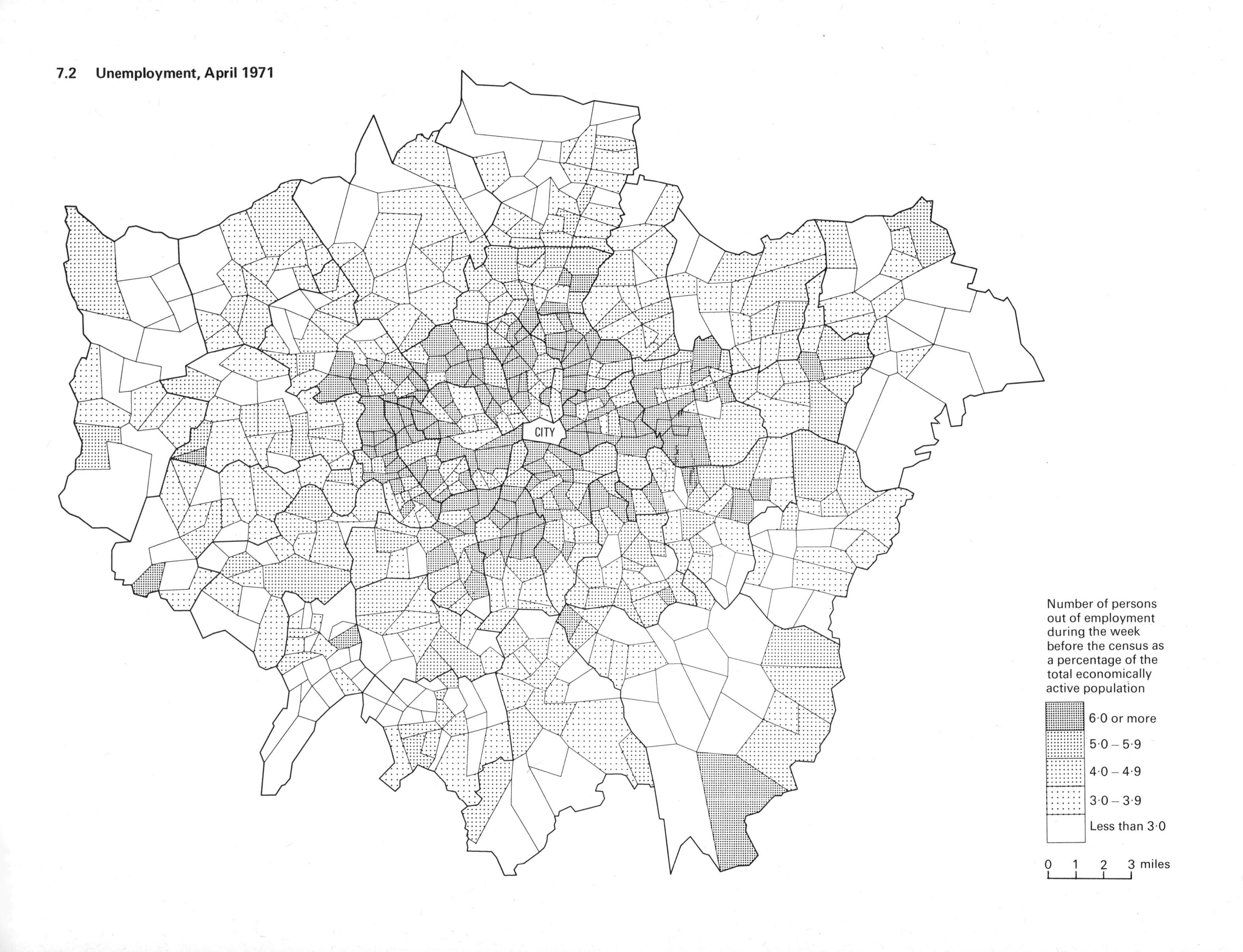

7.3 Achievement in primary education – Inner London 1968

The concept of equal educational opportunity within the State school system has undergone several changes in the past hundred years. A national system of elementary schools was first established in the 1870s. Scholarships were later provided for able working-class children to continue their education beyond the elementary stage. Then, under the Education Act of 1944, children were allocated to grammar, technical, or secondary modern schools according to tests measuring certain abilities and aptitudes; today we are witnessing the gradual replacement of selectivity by various schemes of comprehensive education.

It has been increasingly realized, however, that changes in the formal structure of education have done comparatively little to eliminate inequalities in education. For example, research has shown conclusively that children of unskilled manual workers generally perform less well in school than children of skilled manual and non-manual workers, even when they are of similar measured ability. But social class categories only represent a convenient way of reflecting a variety of factors that have a more direct bearing on a child's educational performance. Children of unskilled workers are more likely to come from large families where less attention can be paid to their individual needs; unless they live in a council house they are more likely to live in over-crowded conditions; there will probably be fewer books in the home; their parents are less likely to take an active interest in their education; and they are less likely to have had any 'pre-school' experience such as nursery school or play group. Thus, inequalities of educational opportunity have come to be recognized not only in terms of a child's ability but also in terms of inequalities in the *total environment* in which he or she develops, including the home, the school, and the neighbourhood.

Map 7.3 represents a generalization of the outcome of these, and other factors in terms of educational attainment in the primary schools of Inner London in 1968. A school's level of educational attainment was measured by the proportion of its children who scored in the bottom twenty-five per cent of all Inner London children on at least one of three tests in English, Mathematics, and Verbal Reasoning. The tests were taken by children in their final year at primary school and the results were standardized for the I.L.E.A. area as a whole. A choropleth map of educational attainment on a ward[1] basis was derived by first assigning schools to wards according to the extent to which a notional catchment area of a quarter of a mile around each school was contained within a ward, and then averaging the test scores for the schools involved. Where a school catchment area was equally divided between two or more wards, the school attainment score was allocated to *all* the respective wards.

Even at this relatively high level of generalization the patterns on the map confirm the relationship between social class, environment, and educational performance. Taking wards with an average of more than 30 per cent of children in the low quartile of scores as areas of low attainment, and those with less than 20 per cent of children in the low quartile as areas of high attainment, the main geographical contrast in educational performance emerges between the predominantly working-class, high-density areas immediately surrounding the centre, and the middle-class, more suburban areas on the fringes of Inner London. Among the low attainment areas are most parts of Islington and Hackney; Bethnal Green, Bow and Poplar in Tower Hamlets; Deptford (Lewisham), Walworth, Camberwell and Peckham (Southwark), Brixton (Lambeth), Battersea (Wandsworth), and parts of Hammersmith, Kensington, and Westminster. Among the high attainment areas are Streatham Vale and West Norwood (Lambeth), Dulwich (Southwark), and Kidbrooke and Eltham (Greenwich). Central Westminster and South Kensington, areas of high social status, also stand out as areas of high educational attainment. Finally, note the scattered nature of areas of the very lowest educational performance (over 40 per cent of children in the lowest quartile), and the way in which they often contrast markedly with the average performance of contiguous areas.

In an effort to counteract the debilitating effects of a bad environment on educational performance the Plowden Report on *Children and their Primary Schools* (1967) recommended that additional resources be made available to schools in objectively-defined *education priority areas* within the inner city. This 'positive discrimination' in the allocation of resources was needed because schools in slum areas had two tasks to perform: to provide the right environment for modern education *and* to compensate for the deprived homes and neighbourhoods from which their children came. The Inner London Education Authority subsequently devised an e.p.a. indicator based on the social class composition and housing conditions in a school's catchement area, together with measures of absenteeism, pupil and teacher turnover, proportion of immigrants and the 'up-take' of free meals within each primary school. In this way over 180 out of 876 primary schools in Inner London were designated e.p.a. schools in 1968.

[1] 1966 ward boundaries were used in compiling this map.

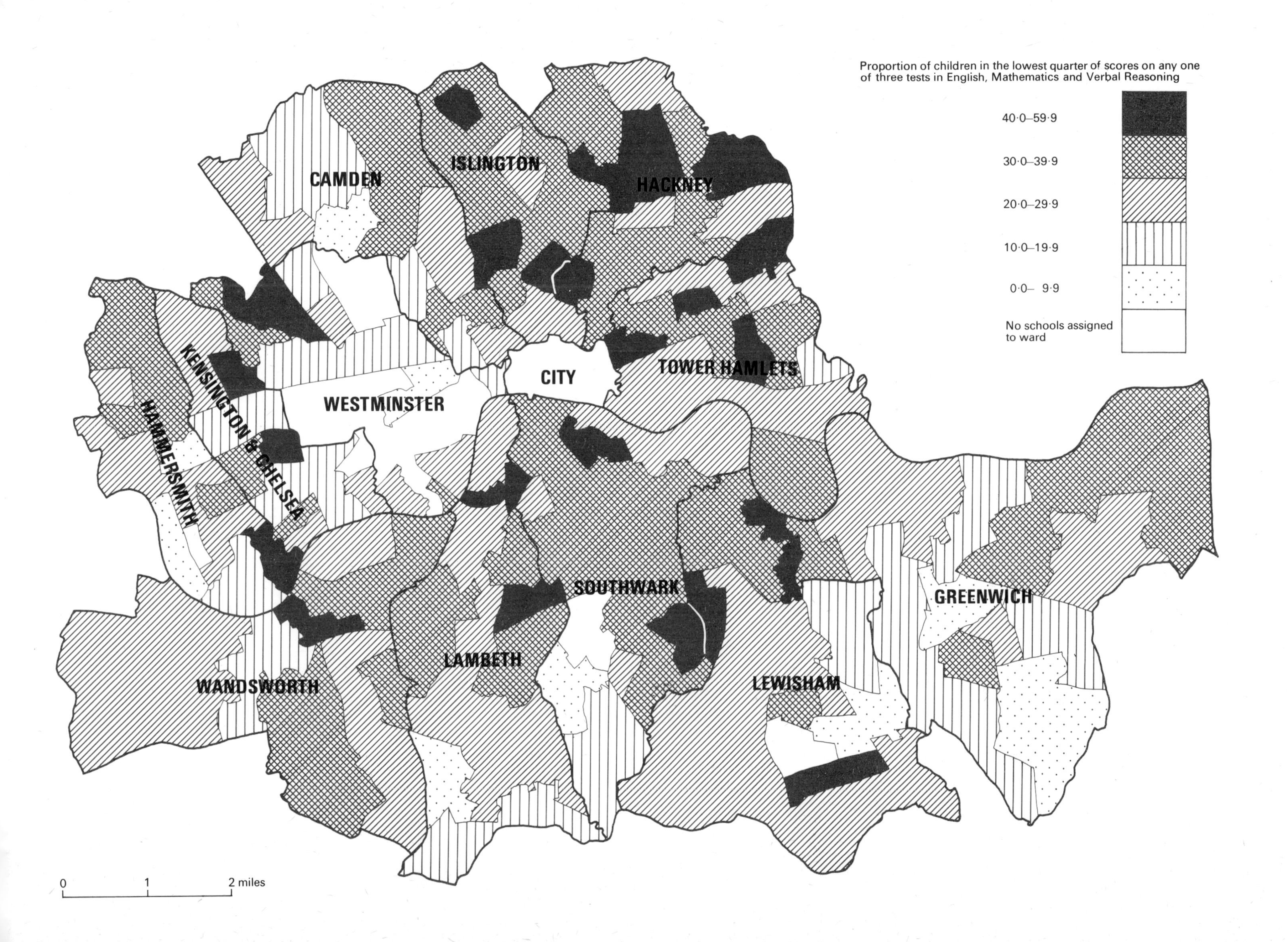

Proportion of children in the lowest quarter of scores on any one of three tests in English, Mathematics and Verbal Reasoning
40·0–59·9
30·0–39·9
20·0–29·9
10·0–19·9
0·0– 9·9
No schools assigned to ward
CAMDEN
ISLINGTON
HACKNEY
KENSINGTON & CHELSEA
HAMMERSMITH
CITY
TOWER HAMLETS
WESTMINSTER
SOUTHWARK
GREENWICH
LAMBETH
WANDSWORTH
LEWISHAM
0
1
2 miles

7.4 Educational provision

We have isolated three components (see Appendix 2) for education in London. The first is *awards for education and staying at school* and it consists of variables measuring success in climbing the education ladder from primary school to college or university. There are many factors which can determine a child's chances of reaching the top of the ladder; his or her potential ability, parental attitudes to education, quality of primary and secondary schooling, and opportunities in higher education.

The importance attached to potential or innate educational ability lies at the root of the controversy over selective versus comprehensive secondary education. Intelligence tests taken at the age of eleven attempt to measure inherent educational ability although it has been shown that the results of such tests are strongly influenced by the environment in which the child develops, the family background, and the social composition of the primary school. It has also been demonstrated that parental attitudes to education in turn depend on social class and income, on the number of children in the family, and on attitudes to marriage and careers for women. The standards of primary and secondary education available in a particular neighbourhood are also very important, especially the supply and quality of teachers and conditions of buildings and equipment.

As the map of the 'awards' component shows, in 1971 there was a clear contrast between the middle-class suburbs to the north-west and south, with high levels of educational success, and the more working-class boroughs in the east with much lower levels of success. We should also note that the stronger the tendency in a borough for men to obtain more awards than women, the lower the overall proportion of awards granted by that borough, so that the boroughs with low success rates are also likely to be the least favourable for women. The situation for Inner London boroughs is one of intermediate, tending to low, success, though the detailed pattern here is blurred by the fact that measures on the characteristics comprising this component are only available for the Inner London Education Authority as a whole. For the purpose of this analysis therefore each inner borough had to be assigned the overall value for the I.L.E.A.

Component two, *primary schools*, distinguishes boroughs with more children per school (the range is from 381 in Harrow to 197 in Richmond), and larger numbers of pupils per teacher (28·4 in Bromley to 20·4 in Islington), from those with relatively high proportions of children in denominational schools. Before 1870 the churches provided almost all elementary education but, since 1900, the number of religious schools has steadily declined as the State has demanded higher standards of education and the cost of building new schools has increased. Hence the boroughs which still have many church schools are those which were substantially built-up by the late nineteenth century. This early urban development is reflected in the scores for suburban Richmond and Kingston while, in Inner London in such relatively old boroughs like Camden, Westminster, and Hammersmith there is, at the same time, continuing support for Roman Catholic Schools from the resident Irish population. In contrast to these 'denominational' boroughs are those in the suburbs with more pupils per primary school and higher pupil to teacher ratios. High scores on this measure do not necessarily indicate that schools in these boroughs are overcrowded. It may instead mean that newly-built primary schools are designed to accomodate more pupils and that, in boroughs where the school population is growing rapidly, the demand for school places is increasing faster than the recruitment of teachers.

The third component, *secondary education,* differentiates between boroughs which still have a form of selection for grammar, technical, or secondary modern schools and those which have almost achieved fully comprehensive education. Note, too, that the proportion of immigrant pupils in secondary schools is related to the proportion of pupils in comprehensive schools. The geographical distribution of this component is not simple, but reflects the complexity of factors involved in the decision to change to a comprehensive system. The political outlook of the education authority is one very important factor. Thus the London County Council was a Labour-controlled authority and a pioneer in comprehensive education. Its policies of educational reform have been continued by the I.L.E.A., which is also dominated by the Labour Party. But, even within Inner London, the pace of change to comprehensives has been partly determined by the availability of finance for purpose-built schools and by such local, physical circumstances as accessibility between existing schools which might be joined in a comprehensive scheme. Labour boroughs outside Inner London which pressed ahead with comprehensive education in the mid 1960s include Brent and Barking, while Newham and Hillingdon maintain selective systems (as at 1972). Of the Conservative-controlled suburban boroughs Harrow, Richmond, and Bromley have been consistently against comprehensive education, Enfield adopted a full comprehensive system, and Merton and Kingston have established one of the permitted versions of comprehensive organization laid down in the Government circular on the subject in 1965.

Maps 7.4—7.6 were designed by using principal components analysis, a statistical technique which is explained in Appendix 2. The commentaries to the maps assume a general understanding of the method.

7.4 Components of educational provision

(a)

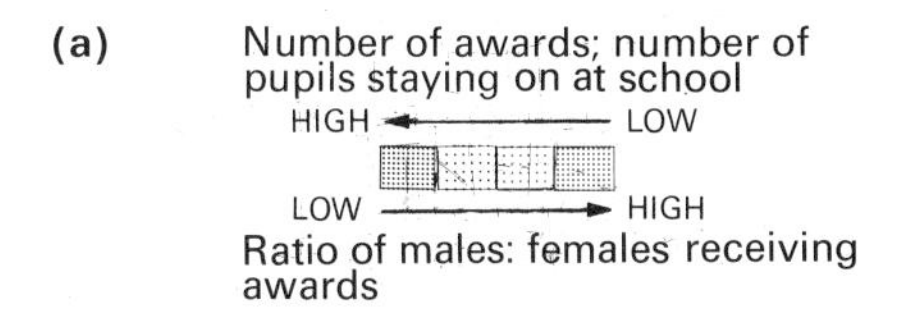

(b)

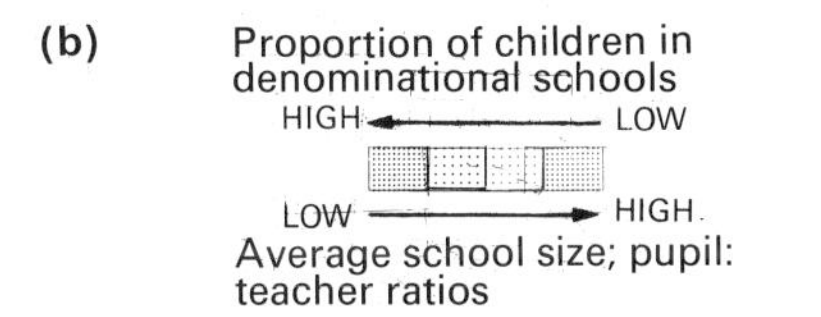

(c)

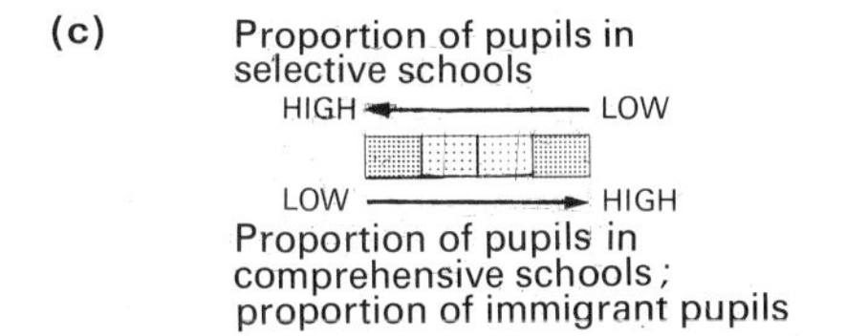

0 1 2 3 4 5 miles

(a) Awards and staying on at school

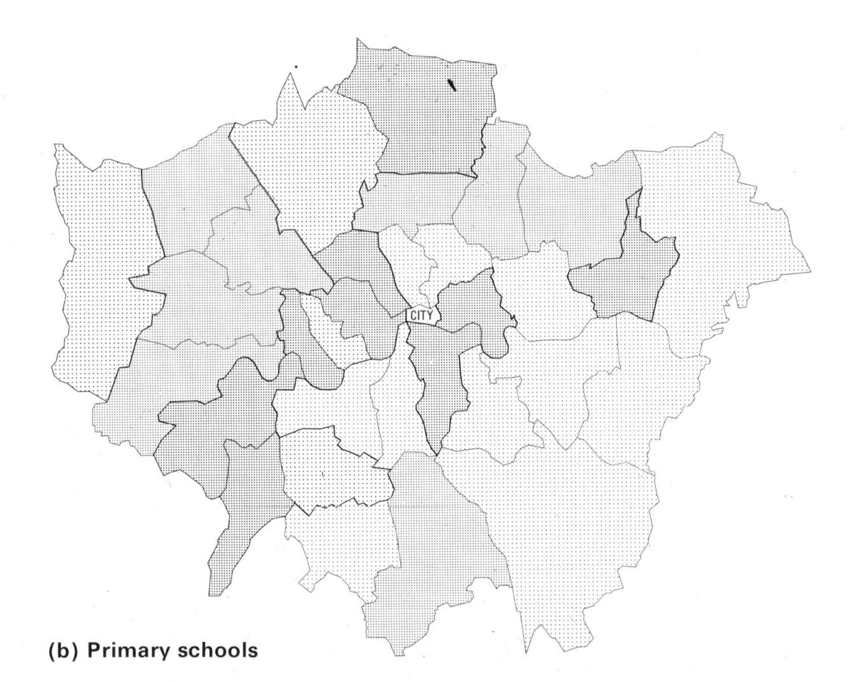

(b) Primary schools

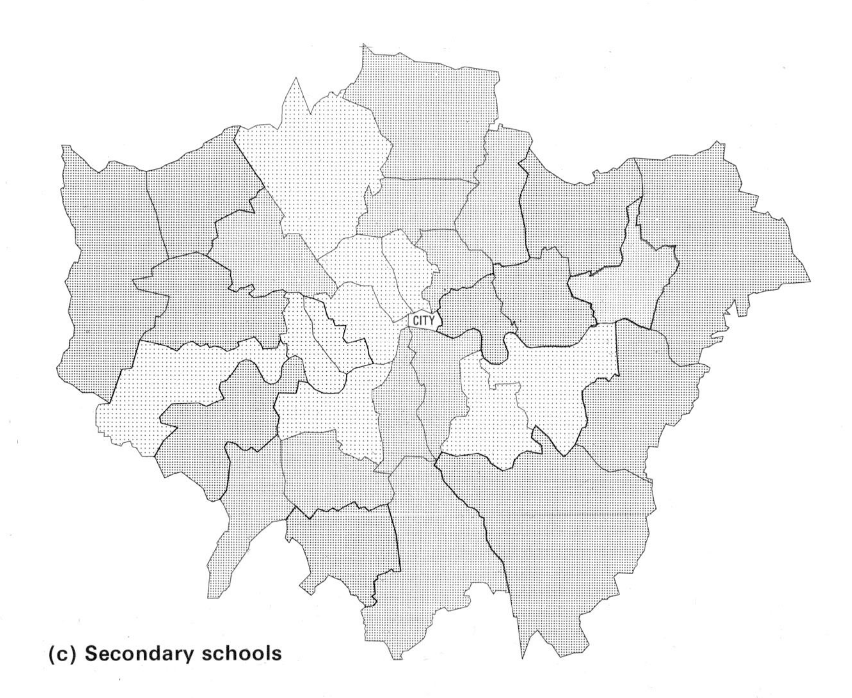

(c) Secondary schools

7.5 Metropolitan environment

Up to this point most of the maps in this atlas have been concerned with a single topic in its spatial and social context. For example, in the section on transport and shopping it was seen that the higher incidence of road accidents in inner London can be linked to the social composition of the area and to features of the physical environment which expose more people to the risks of traffic. In other words, each particular topic is related to something which we might loosely call the *urban environment*, but which itself is an intangible concept embracing all the specific relationships which make up the distinctive character of an area. The three maps opposite represent some preliminary indicators of 'urban environment', which emerge from a study of a given set of variables (listed in the appendix), measured for each of the London boroughs by using principal components analysis.

The first component is characterized by the strong interrelationship between seven of the original fifteen variables that were included in the analysis. Five of them—population density, mileage of principal roads in relation to residential land, deaths from road accidents, and smoke and sulpher di-oxide poisoning in the air—are all indicators or consequences of an intense concentration of people and activity. The other two—illegitimate birth rate and rate of referral of mentally-ill people to the local authority—can be seen as evidence of social breakdown, and the loneliness and anonymity associated with the inner areas of big cities.

The second component, *cancer and circulatory diseases*, brings together three variables all bearing the same relationship to each other. Two of them—deaths from heart and other circulatory diseases and deaths from cancer other than of the lungs—we expect to be related on medical and, perhaps, social grounds. The third variable is the amount of open space per head of population in a borough. The component appears to suggest that, the more open space there is, the higher the incidence of heart disease! Such a relationship has either occurred by chance, or there are intervening variables missing from the analysis. We might argue, for example that the open-space variable really measures the quality of 'suburban-ness' and hence 'middle-class-ness'. But the geographical pattern of the component offers little help in explaining what such linking variables, if any, there might be and we must clearly beware of inferring too much from the relationships. There are high scores on this component in both inner and outer boroughs; in boroughs of high social class like Kensington and in working-class boroughs like Tower Hamlets; and in boroughs with both high and low proportions of elderly people, like Enfield and Richmond respectively.

The third component, *air-pollution, living conditions, and respiratory diseases,* offers much more scope for interpretation. Like the first it includes both measures of air-pollution but, in this case, it singles out those boroughs in which higher-than-average concentrations of smoke and sulpher di-oxide in the air occur *in conjunction with* respiratory diseases, bad housing conditions, and mortality at birth. It therefore brings together a group of variables in an almost classic syndrome typifying the unhealthy urban environment.

The geographical expression of this component is arresting. There is a continuous arc of very low scoring boroughs in the west of London stretching from Barnet in the north, through Hillingdon, Richmond, and Kingston to Sutton in the south. High scores on the component are found in a large bloc of eastern boroughs including suburban Waltham Forest and Redbridge. There is an irony here for, although eastern boroughs like Lewisham, Greenwich, Tower Hamlets, Hackney, and Redbridge (but not Newham) have brought a much higher proportion of premises under Smoke Control Orders (Clean Air Act 1956) than the western boroughs, London's prevailing westerly wind blows much of the pollution of the western boroughs into the atmosphere of the eastern boroughs.

7.5 Components of metropolitan environment

(a) Road accidents; population density; illegitimate births, air pollution mental health referrals
LOW → HIGH

(b) Death rates from certain cancers and circulatory diseases; open space
LOW → HIGH

(c) Proportion of poor or unfit dwellings; air pollution; deaths from respiratory diseases
LOW → HIGH

0 1 2 3 4 5 miles

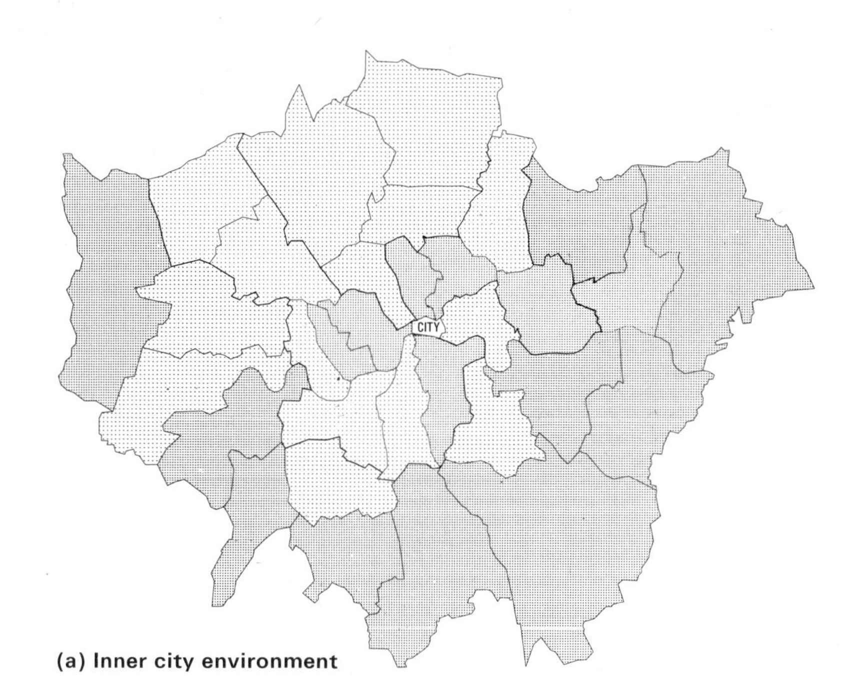

(a) Inner city environment

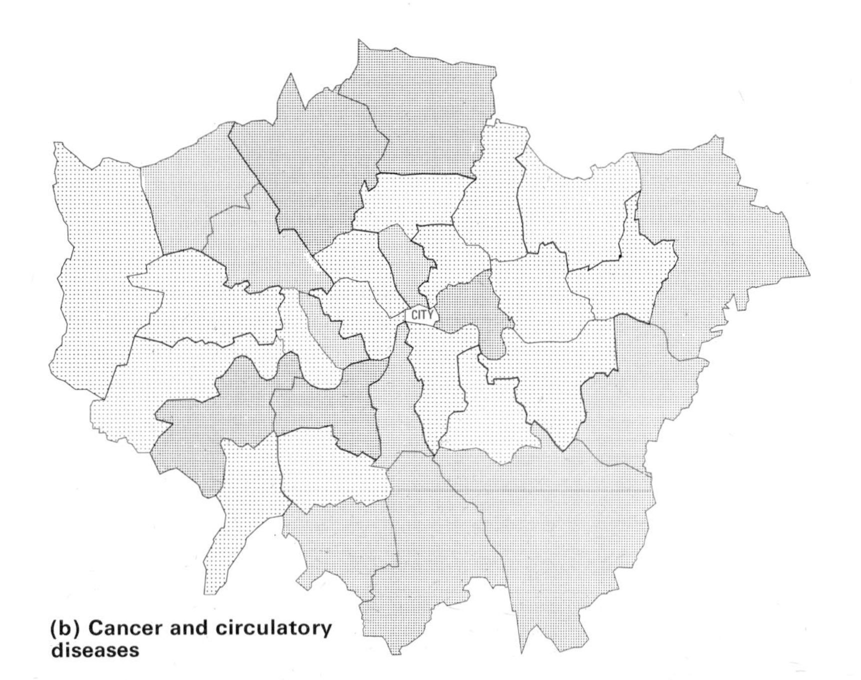

(b) Cancer and circulatory diseases

(c) Air pollution, living conditions and respiratory diseases

7.6 Borough expenditure on planning and welfare

During 1970-71 it cost £1593 million to keep London's essential services running, an increase of £250 million over the previous year. The authorities responsible for this expenditure are the G.L.C. and I.L.E.A., the London boroughs, the Metropolitan Police, and the Metropolitan Water Board.

The greater part of the total annual spending (£1140 million in 1970-71) is *current* or *revenue* expenditure which, in paying for salaries, supplies, and maintenance, covers the recurring costs of providing services. Revenue expenditure is paid for largely out of the rates and central government grants. The remainder of the total was *capital* expenditure which comprises investment in new buildings, facilities, and fixed equipment. Spending by the G.L.C./I.L.E.A. and London boroughs alone amounted to £1465 million in 1970-71, of which the G.L.C. share was £501 million and that of the boroughs £964 million. The most important categories of expenditure were education (£346 million), housing (£211 million), and health, welfare, and child-care (£96 million). Revenue expenditure amounted to almost three-quarters of total spending by the boroughs. There are, however, marked variations in expenditure between the London boroughs, as is shown on the two component maps opposite.[1]

These two components distinguish between revenue expenditure per head of population on a group of *social welfare* services, and on two *environmental* services. The geographic pattern of the first component displays the familiar contrast between inner and outer London, with the inner boroughs generally spending much more per head of population on social welfare. Note, however, that Newham, a borough with many of the social and environmental characteristics of Inner London, tends to spend relatively less on this particular group of services. The six boroughs which spend least per head of population on social welfare are all suburban.

The second component of expenditure is significant in that it reveals those inner boroughs which also spend relatively more on the two environmental services. These are Camden and Westminster in particular, and Hammersmith and Islington to a lesser extent. Kensington and Chelsea and Wandsworth, which are also Inner London boroughs, are more like the group of eastern boroughs and Brent in the west, which have the lowest 'scores' on this component.

The determinants of the level of local authority spending can be grouped into three sets of factors. There is first the extent of need in an area. As we have seen in commentaries to other maps, *need* is related to the social-class composition of a borough, its rate of population change, and the age and sex structure of the population. Differences in need arising out of these characteristics largely account for the great contrast between inner and outer boroughs in expenditure on health, welfare, and child-care services. Next, there are the *conditions* under which a service is provided which affect the cost of provision. The population size and composition of a borough will determine whether there is a sufficient caseload to warrant employing specialists in a particular service or establishing a separate department where the authority is under no statutory duty to do so, and the physical character of a borough including the size of its area, population density, communication patterns, even its shape, will affect the time and expense of travelling undertaken by social workers, health visitors, and home-helps. Finally, there are factors of *politics and personality* which affect the emphasis placed on some services compared to others. Included here are the party-political composition of the council, the way politicians perceive need and are influenced by public opinion, and the advice given to them by senior officers.

We should also note that, just as there are differences between the boroughs in expenditure, so there are contrasts between them in the amount of money they can raise through the rates. Over one-third of Greater London's total rateable value of £680 million in 1971 was concentrated in only four authorities, namely Kensington and Chelsea, Westminster, Camden, and the City of London. These differences are reduced to some extent by central government which makes grants to local authorities on the basis of need and population, and there is also a complicated rate equalization scheme in London, whereby richer authorities contribute to the costs of poorer ones. But there are glaring anomalies appearing in both these attempts at equalizing the distribution of finanacial resources between boroughs as shifts take place in their population levels and in their residential, commercial, and industrial character.

[1] Two points should be made about the expenditure figures. First, they represent sums over which councils have a considerable degree of control. Second, they cannot tell us whether the boroughs are meeting all the demand or need for services in their area or whether they are getting value for money. For example, high rates of expenditure per head of population on child care may mean that a borough is providing both an *effective* service in terms of seeking out and rehabilitating neglected children, and an *efficient* one in terms of cost to the ratepayers. But high spending could also imply that a borough is providing inadequate service at high cost.

7.6 Components of revenue expenditure

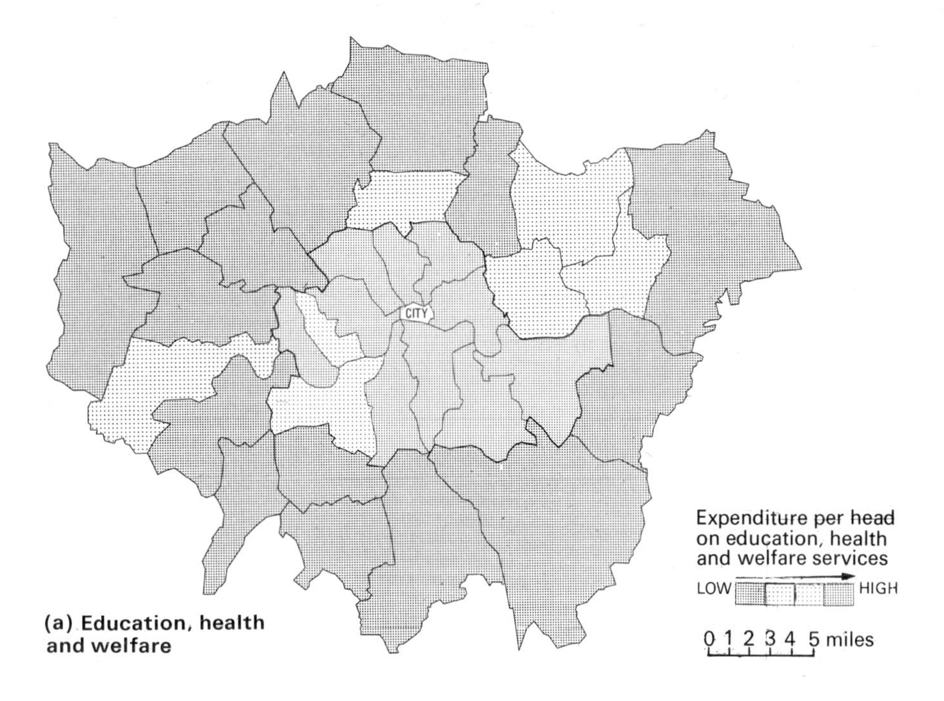

(a) Education, health and welfare

(b) Town planning and open space management

Local elections

8.1 Dividing a London borough into wards

The wards of Greater London are not simply areas abstracted for the presentation of census statistics. They have a social significance of their own as being, first and foremost, the territorial basis for representation on the London borough councils. Any proposal to change the boundaries of wards, their size, shape, and configuration is therefore of considerable interest to individual politicians and to the major political parties. Sometimes the conflict over re-warding is such that a public inquiry is held to determine a particular scheme of sub-division and it is in this inquiry that the influence of the big parties on re-warding is revealed.

In preparation for the changeover to the new system of local government in 1965 the constituent councils of the new London borough of Westminster were invited by the Home Office to submit proposals for re-warding their area and apportioning a maximum of 60 councillors among the new wards. Central government gave only very broad criteria for re-warding. Account had to be taken of short-term changes in the distribution of population; the ratio of councillors to electors was to be about the same in every ward; ward boundaries should be easily identifiable and, as far as possible, local community ties should be respected in fixing a particular boundary. No guidance was given as to why these four criteria were important or how to measure and weight them in the resolution of the political battles that would accompany the re-warding process.

The new borough of Westminster, which would clearly have a Conservative majority on the council, was formed by amalgamation of the former boroughs of Paddington, St. Marylebone, and Westminster. Agreement on the re-warding of Paddington and St. Marylebone was reached by Labour and Conservatives with little difficulty. Labour had very little chance of winning a seat in the St. Marylebone area while in Paddington its quite considerable support was *concentrated* in three contiguous wards of working class and immigrant settlement, namely Queens Park, Harrow Road, and Westbourne. The proposed amalgamation of wards in this area (compare Maps a and d) would have had little effect on the total number of seats that the party might win. It was dissension over the re-warding of the former Westminster area that led to a public inquiry into the issue in late 1963.

In the southern part of the new borough the Labour Party was still electorally overwhelmed in aggregate but it did have scattered pockets of traditional support which, in a good year for the party, could bring it five extra seats on the council. These Labour enclaves were in the small Tachbrook, Millbank, Ebury, Churchill, and Soho wards. The Conservative proposal would have created five very large wards in Westminster (Map b), each enveloping one or two of these enclaves by 'swamping' Labour supporters with Conservative votes from neighbouring wards. The Conservative case for larger wards was that they were in line with the general 'scaling up' of local government areas under the London Government Act. Also, as each ward would have four representatives, there would be less disruption due to changes in the size of electorate or the death or retirement of a sitting member than with small, single-councillor wards. In addition, large wards with several members had a greater chance of at least some representation on a council committee.

The Labour Party objected strongly to this amalgamation of wards and proposed 20 'single-member' wards instead (Map c). Small wards, it was considered, made for closer contact between councillor and electorate, generated greater interest in local government and, as there would be fewer names on the ballot paper, caused less confusion and made issues clearer for electors. According to the Labour spokesman, successful political representation depended on a community of interests in common problems and this was more likely to be achieved in small, socially homogeneous wards. Inhabitants of a ward which, as the Conservatives proposed, could combine high-class shopping and residential areas like Mayfair and St. James's with entertainment and more working-class areas like Soho and Covent Garden, would not share problems in common.

Having considered the arguments outlined above, the Commissioner at the inquiry eventually pronounced in Labour's favour, though the force of Labour's case was probably strengthened by the fact that it would cause less disruption to existing ward boundaries and could be acted upon within the six months before the first elections to the new London boroughs took place. Later the re-warding issue arose twice more in the London borough of Westminster. The Paddington area of the borough was re-warded in 1966 and the old Westminster part in 1971. On the latter occasion the Labour opposition on council accepted a compromise of medium-sized wards (Map d) rather than fighting the issue in public again.

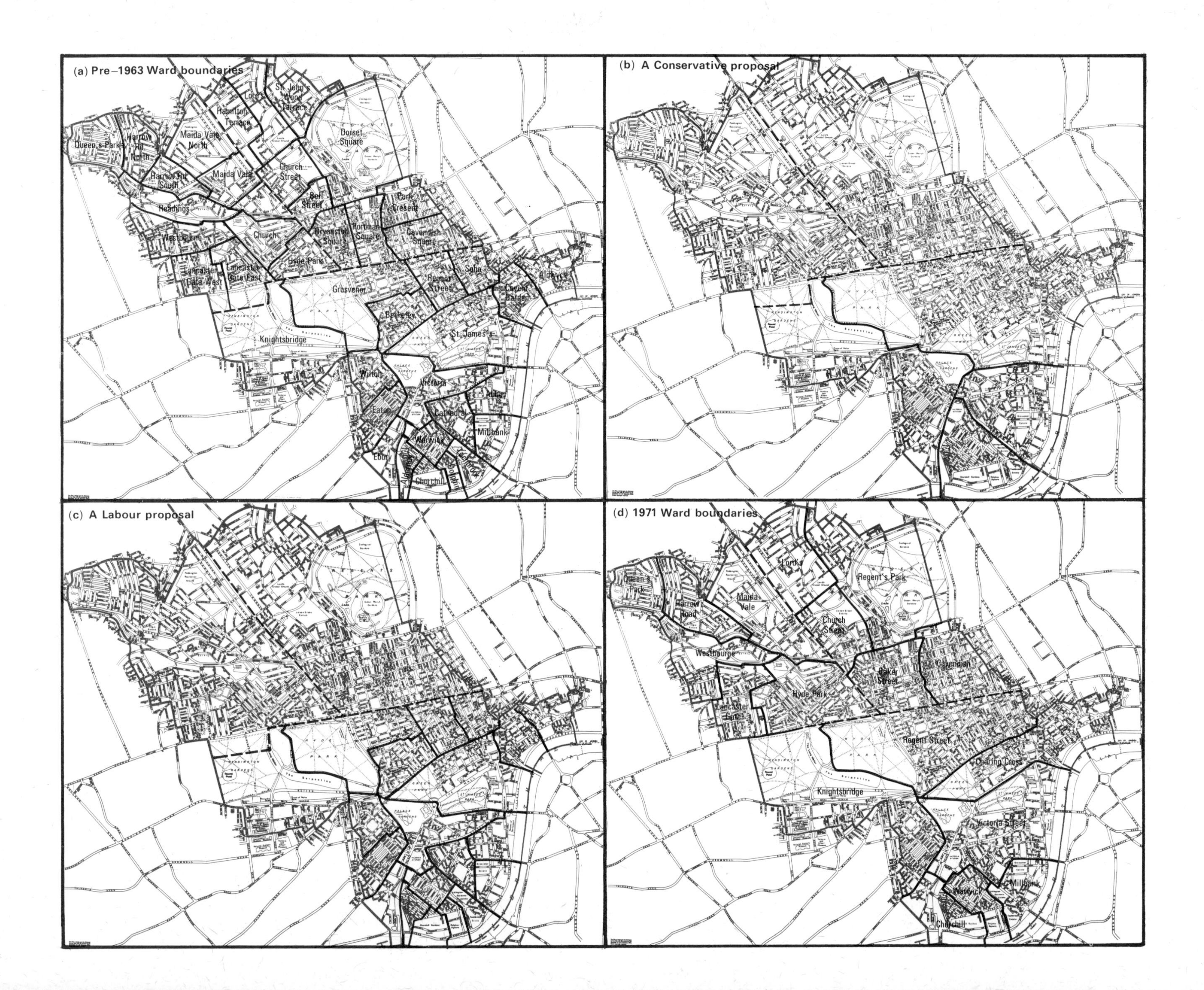

(a) Pre–1963 Ward boundaries
Queen's Park
Harrow Rd North
Harrow Rd South
Maida Vale North
Maida Vale
Hamilton Terrace
Lords
St. John's Wood Terrace
Dorset Square
Church Street
Bell Street
Park Cresent
Readings
Westbourne
Church
Bryanston Square
Portman Square
Cavendish Square
Hyde Park
Lancaster Gate West
Lancaster Gate East
Grosvenor
Regent Street
Soho
Aldwych
Covent Garden
Berkeley
St. James's
Knightsbridge
Wilton
Victoria
Abbey
Eaton
Cathedral
Millbank
Warwick
Ebury
Churchill
(b) A Conservative proposal
(c) A Labour proposal
(d) 1971 Ward boundaries
Queen's Park
Harrow Road
Maida Vale
Lords
Regent's Park
Church Street
Westbourne
Baker Street
Cavendish
Hyde Park
Lancaster Gate
Regent Street
Charing Cross
Knightsbridge
Victoria Street
Warwick
Millbank
Churchill

8.2 London borough elections 1971: Turnout

Electoral arrangements

London borough elections are held every three years and all the seats on the thirty-two councils are contested at the same time. The first borough elections were held in the spring of 1964, though the second and third took place in 1968 and 1971, following an act of parliament which delayed by one year the elections which were to have taken place in 1967.

The London borough councils have from 48 to 60 councillors, although each council includes aldermen in addition, numbering up to one-sixth of the council. Aldermen are elected by the councillors and, as happened in Harrow in 1971, when the outcome of an election leaves no political party in absolute control, the choice of aldermen can be decisive.

Although twenty-four boroughs have adopted the maximum number of sixty councillors permitted by law, the method of allocation of councillors to wards varies considerably. Only nine boroughs have a uniform number of councillors to wards, usually three members per ward, though Enfield has a two-member system. Most other boroughs have adopted a combination of two, three, four, or five members per ward. However, before its ward boundaries were changed in 1973, Kensington and Chelsea had every number of members per ward from two through to seven in a total of only fifteen wards. Almost half the council then sat for only four wards, namely Earls Court, Pembridge, Redcliffe, and St. Charles. A uniform or near uniform number of councillors per ward does not necessarily imply equality of representation for the individual elector, but it does mean that wards have a roughly equal chance of being represented on an influential committee of the council.

Turnout

We must note that the maps of, for example, housing, population structure, or social class, measured a given *characteristic*, whereas Maps 8.2, 8.3, and 8.4 measure the way people *behave* (when voting in this case). The range of factors, from the psychological to the sociological, involved in human motivation can make the interpretation of such maps difficult.

Percentage turnout in elections is often taken as a measure of public interest in the democratic process. High turnout is assumed to mean considerable interest in an election, low turnout to imply public apathy, and there appears to be some evidence to support this view. The proportion of people voting in general elections is much greater than in local government elections. Turnout for Greater London as a whole in the General Election of 1970 was 64·0 per cent—rather low by the standard of previous elections—compared to 37·0 per cent in the G.L.C. election of 1973 and 38·7 per cent in the 1971 London borough elections.

Evidence linking turnout to level of interest or sense of civic responsibility has also been gathered by means of interview surveys carried out directly in the context of London elections. These have shown that professional, managerial, and other non-manual occupation groups vote more than manual workers and, within the manual occupations, that skilled workers vote more than the unskilled. Owner-occupiers have been found to have a higher rate of turnout than the tenants of private landlords and local councils, as have electors with higher levels of education and income. Aggregate-level information could also be interpreted as supporting this view and the map of electoral turnout by ward reveals the same broad contrast between inner London boroughs, including Newham, and the outer boroughs, that has been shown before on the maps of social class and housing tenancy.

Low turnout at local elections need not, however, be taken as evidence of political apathy; it can also be seen as a passive protest, which does not go so far as voting for the other side. Such protest may be a deliberate act of abstention or it may be an instinctive reaction to a feeling that local government is remote and unresponsive to the situation of the individual. From aggregated information for wards it is impossible to produce direct evidence on such individual motivation but there may be some indirect pointers. Turnout is extremely low in working-class boroughs of inner London like Hackney, Tower Hamlets, and Newham as well as in upper-class Kensington and Chelsea and Westminster and what these boroughs have in common are virtually impregnable majorities for the Labour and the Conservative parties respectively. In boroughs like these there is unlikely to be sufficient political conflict to arouse interest in an election and the individual elector is more likely to feel that his vote will make little difference to the outcome, even in his own ward. On the other hand, turnout in both Hammersmith and Greenwich was very high, though both these boroughs returned overwhelming Labour majorities on their councils in 1971. Part of the explanation may lie in the fact that the Conservatives won both boroughs in 1968 but, nevertheless, although Greenwich is a predominantly working-class area, it has maintained a tradition of relatively high electoral turnout since the creation of the London County Council in 1888.

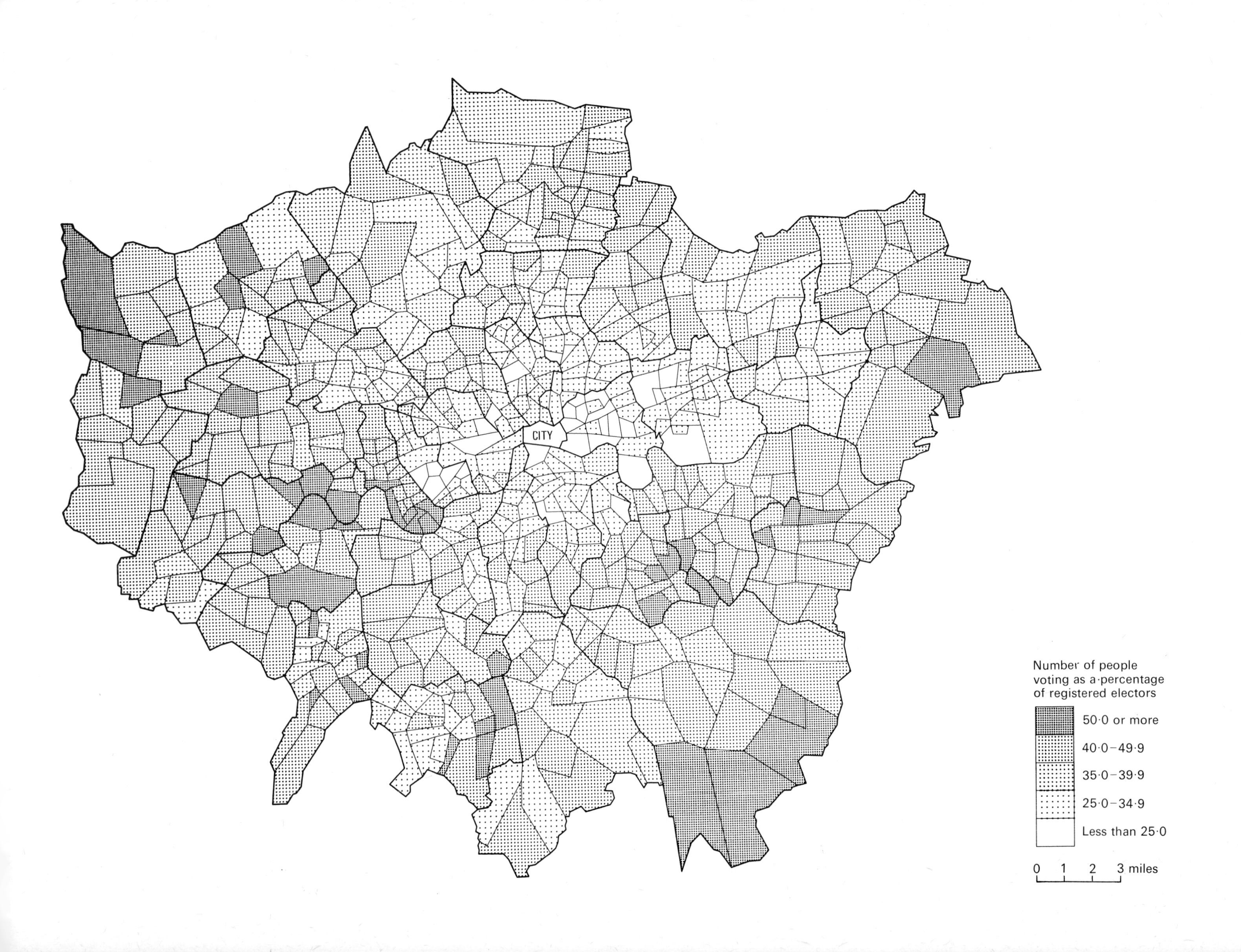
CITY
Number of people voting as a percentage of registered electors
50·0 or more
40·0–49·9
35·0–39·9
25·0–34·9
Less than 25·0
0 1 2 3 miles

8.3 London borough elections 1971: Conservative, Labour, and Liberal vote

The politics of local government in London, as in other urban areas, are dominated by the two big political parties. No less that 3606 of the 4635 people who stood for election to the London borough councils in 1971 were endorsed by the Conservative or Labour parties. The Liberal Party fielded 595 candidates. Because of this identification between national and local politics, the results of local elections are largely determined by the strength of public feeling about the performance of the government at Westminster. Local government candidates may appeal for the support of local voters on such issues as education, housing, public transport, and the rates but, except in certain circumstances, their electoral fate is sealed by matters of national concern like the cost of living and industrial relations.

The results of the first three London borough elections in terms of the number of boroughs controlled and the total number of seats won by the main political parties, are set out in the table below. They reveal the sweeping effects of fluctuations in the popularity of the party in control of the central government. In 1964 the London borough elections took place six months before Labour narrowly won the October general election. The elections of 1968 virtually coincided with the lowest level of popularity of the second Labour Government and those of 1971 occurred almost a year after the return to power of the Conservatives.

Number of boroughs and seats won in London borough elections

	1964		1968		1971	
	Boroughs	*Seats*	*Boroughs*	*Seats*	*Boroughs*	*Seats*
Conservative	12	667	28	1438	10	597
Labour	20	1112	4	350	22	1221
Liberal	–	16	–	10	–	9
Others	–	64	–	65	–	36

In 1971 Labour Party candidates won 65·5 per cent of the total number of seats with a 53·1 per cent share of the total votes cast. However, such figures conceal a considerable variation between the boroughs in the number of councillors per ward, the number of electors in a ward, and the way ward boundaries are drawn. One result of this variety of electoral arrangements is that, for a given election there are differences between the boroughs in the value of a vote to a particular

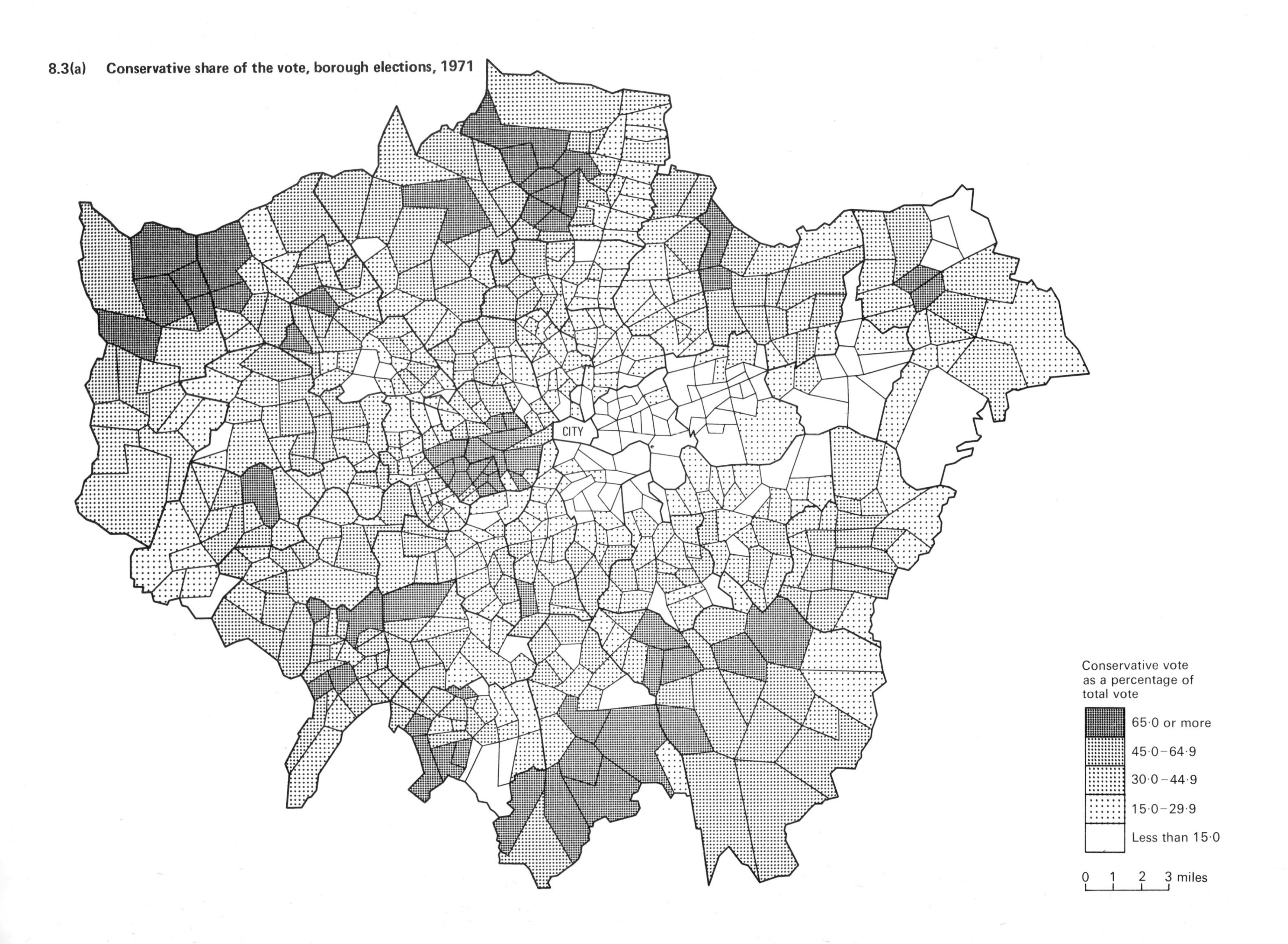

8.3(a) **Conservative share of the vote, borough elections, 1971**

party. In Richmond in 1971, for example, the Conservatives received only 45 per cent of the vote, yet this was enough for them to obtain 74 per cent of the seats on the council. In Barnet less than half of the vote in the borough won the Conservatives 72 per cent of the seats. Among the Labour held boroughs, 70 per cent of the vote in Greenwich won 92 per cent of the seats, and 62 per cent of the vote in Hammersmith won all but two of the sixty seats on the council. The result in the 1971 election of the re-warding of Westminster was that 51 per cent of the votes cast won the Conservatives 62 per cent of the seats.

The maps of the percentage of the votes cast in each ward going to the Conservative, Labour, and Liberal parties in the 1971 borough elections are of interest for two reasons. First, they reveal, albeit for one election only, the extent to which the spatial patterns of support for the two major parties are complementary; that is they show in detail their respective areas of dependable and solid support. Secondly, they suggest where the contest between Conservative and Labour makes the outcome of an election 'marginal' and exactly where, and with what effect, the Liberals have been challenging the established positions of the two main parties.

The relationship between upper- and middle-class social status, relative affluence, and conservatism is brought out by the concentration of support for Conservative Party candidates in the suburbs, particularly those of the north, south-west and south, and in Inner London, in Kensington and Chelsea and Westminster. Except for the latter two boroughs, those that the Conservatives won outright in 1971 were all on the outskirts of London, namely Redbridge, Enfield, and Barnet to the north and Richmond, Kingston, Sutton, Croydon, and Bromley to the south. Within this broad sweep of Tory suburbs there were individual wards which, even at a time of general dissatisfaction with the Conservatives, recorded high levels of support for the party. In Winchmore Hill (Enfield), Conservative candidates received 76·6 per cent of the vote, in Shortlands (Bromley), 81·2 per cent, and in Cheam South (Sutton), 88·7 per cent. The highest Conservative shares of the ward vote were, however, found at the centre of London, in Knightsbridge (90·2 per cent), and Brompton (92·8 per cent), which can be explained in terms of their social-class and age/sex characteristics.

The traditional Labour vote is to be found in the working-class and industrial areas of London. Labour's territory of solid support stretches in a wide band along both sides of the Thames from Erith

8.3(b) Labour share of the vote, borough elections, 1971

(Bexley) and Dagenham (Barking) in the east to the boundaries of Tower Hamlets and Southwark where they meet the City of London. A spur of Labour support leads northwards through Islington and Hackney, along the Lea Valley as far as Edmonton and Enfield. On the south side of the Thames the line continues into Wandsworth and Battersea, crossing the river into Hammersmith, and ending in a substantial outlying area of support in Willesden and Kilburn (Brent). A notable feature of the Labour map is the number of more-or-less isolated wards in the inter- and post-war suburbs of London which record very heavy Labour voting. Examples include Carshalton St. Helier (Sutton), Addington (Croydon), Mottingham and St. Paul's Cray (Bromley), Gooshays and Hilldene (Havering), and Hainault (Redbridge). In most cases these are 'out-county' estates built by the old London County Council in conjunction with its slum-clearance schemes. They provide a graphic illustration of why certain Conservative-controlled suburban boroughs have been slow to provide land for current G.L.C. plans to re-house 'inner Londoners'.

It can also be seen from the Labour map that the main area of Labour support is flanked by a large number of wards which, on a detailed inspection of the voting results, would probably be revealed to be 'marginals' in that it would take relatively few votes to change their political representation. The marginal territory is roughly marked by those wards having a Labour vote of between 35 and 55 per cent. These wards are generally located in the inner, Victorian suburbs with mixed social-class composition and a variety of housing-tenancy groups as in Lewisham, Dulwich, Norwood, Acton, and Hampstead, or they are to be found in the inter-war suburbs of Bexley, Sutton, and Redbridge. As marginals they are, or should be, the areas of the most intensive party-political canvassing in an election.

One possible effect of marginality has already been noted, namely that it may increase turnout at an election. A livening of public interest has also been attributed to the intervention of 'third parties'. A comparison of the map of turnout with that for Liberal votes does, in fact, reveal a similarity of pattern in certain boroughs, especially Bromley, Kingston, Richmond, Harrow, and Havering, although in order to be sure that the intervention of the Liberals does increase turnout, we would have to attempt to disentangle the effect of other influences which affect voting in the suburbs. There are, nevertheless, several striking aspects of the pattern of Liberal support. First, their share of the vote is usually very small compared to that of the two big parties, except in the Orpington area of Bromley, symbol of the Liberal revival. Secondly, there are large tracts of inner London where the Liberals put up no candidates or received an insignificant share of the vote and, thirdly, the Liberal election effort in 1971 was focussed in suburban boroughs, especially Conservative strongholds like Richmond, Barnet, and Bromley. Finally, the evidence of voting patterns suggests that there were, in 1971, only a few, usually small, pockets of Liberal activity within Inner London.

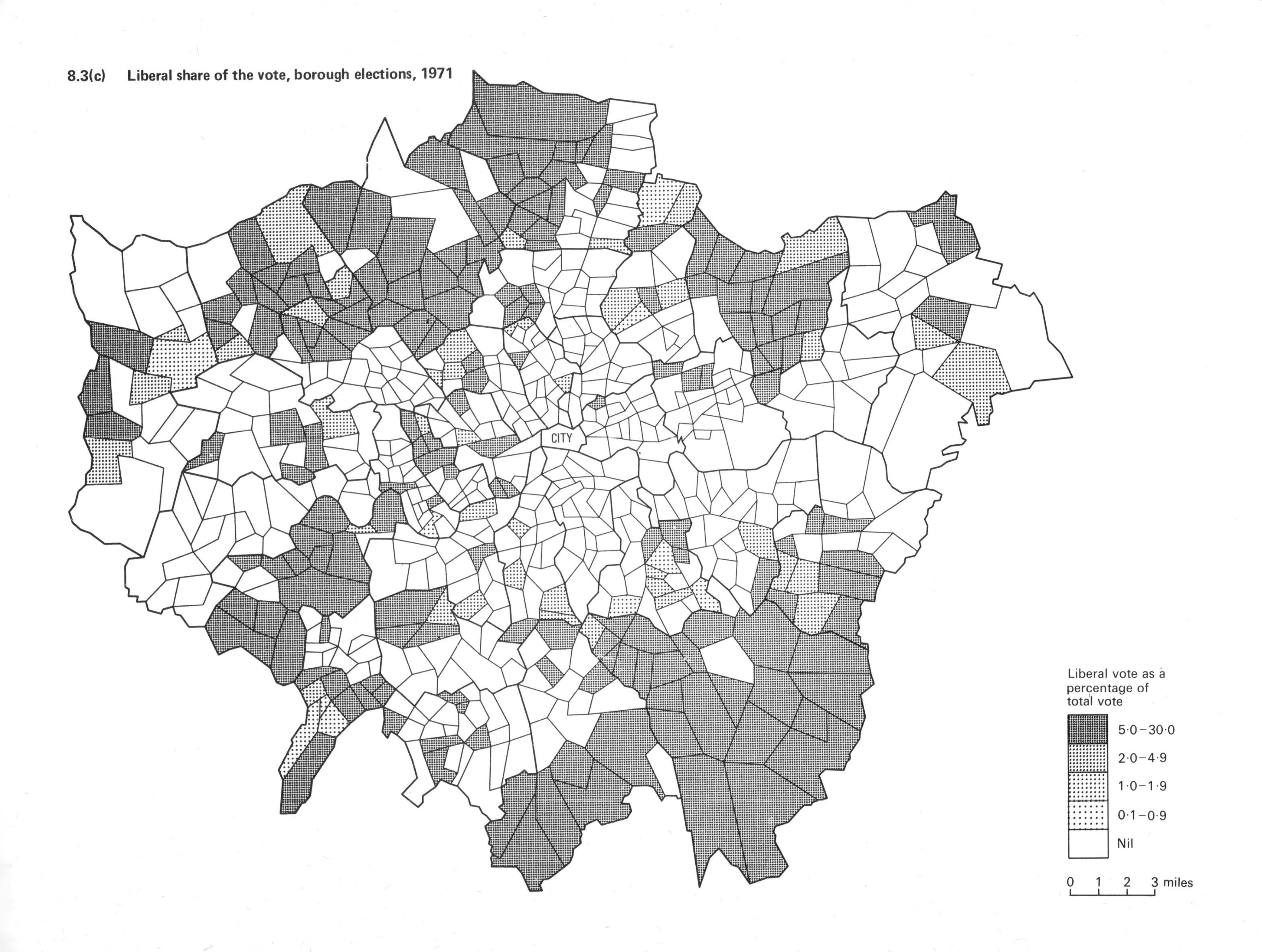

8.3(c) Liberal share of the vote, borough elections, 1971

8.4 Party control and support, GLC elections 1964–73

In the 1973 G.L.C. election some 5·3 million people, roughly one-sixth of the electorate of Britain, were eligible to cast a vote. Turnout in local elections is very much lower than for general elections but, even so, the size and importance of London together with the broad cross-section of social classes and interests that compose its population, mean that the outcome of the party-political contest for the capital is observed closely as an indicator of the mood of the country at large.

Elections to the G.L.C. are held every three years on a date fixed by the Home Secretary. The elections of 1964, 1967, and 1970 were held on a borough basis, each borough electing two, three, or four councillors according to population, and each elector casting one vote per councillor. Under this system there were 100 elected councillors on the G.L.C. and 16 appointed aldermen. The 1973 election, however, brought a change to representation on the basis of new parliamentary constituencies for London, announced in April 1972. There are now 92 'single-member' constituencies for G.L.C. elections each with between 50 000 and 70 000 electors.

One of the effects of this electoral system is that small shifts in public opinion lead to large numbers of seats being transferred between the two main parties (see table). The plunge in Labour support of 11·4 per cent at the 1967 election resulted in a loss of no less than 46 seats on the G.L.C., while a decline of only 2·7 per cent in the Conservative share of the vote in 1970 was enough for 17 seats to change hands. Third parties, like the Liberals, are at a distinct disadvantage under this system. Not even ten per cent of the total vote before 1973 could secure them a seat.

Seats and Votes in G.L.C. Elections

	Conservative		*Labour*		*Liberal*	
	Seats	Per cent vote	Seats	Per cent vote	Seats	Per cent vote
1964	36	41·7	64	46·0	0	10·2
1967	82	54·3	18	34·6	0	10·0
1970	65	51·6	35	40·5	0	5·4
1973	32	38·0	58	47·4	2	12·5

The four maps of G.L.C. elections trace two factors simultaneously. One is the party allegiance of the G.L.C. councillors for boroughs or constituencies after each election. Except in the case of Greenwich in 1967 and Lambeth in 1970, one or other of the main parties won all the seats allotted to a particular borough. The other factor is the strength of support for the two main parties, measured as the percentage of votes cast for them. The categories of percentages used on all four maps are the same, enabling fluctuations in party support to be compared over time.

In terms of the fortunes of the Labour Party, these maps reveal a pattern of fluctuation around a 'core area' of dependable Labour support in inner, or inner suburban areas in east London, namely Islington, Hackney, Tower Hamlets, Newham, Barking, and Southwark. Greenwich should perhaps be added since a by-election held shortly after the 1967 poll left it with two Labour and one Conservative G.L.C. councillors. In the landslide of 1967 Labour not only lost suburban boroughs like Bexley and Havering, which it had succeded in winning in 1964, but also those in the inner London periphery of its core area like Camden, Lewisham, Lambeth, Wandsworth, and Hammersmith. These boroughs were all retrieved in 1970, with the exception of Lambeth, where Labour took one seat and narrowly failed to take the other three. Westminster and Kensington and Chelsea, whose social composition is different from the rest of Inner London, appear throughout the sequence of maps as enclaves of solid Conservative support within the Labour core and periphery.

The change to smaller parliamentary constituencies in 1973 refined the geographical pattern of G.L.C. elections. For example, given that there was a moderately strong swing of public opinion to Labour at this election, the change highlighted extensions of the Labour 'penumbra' in Tottenham, Wood Green, Erith and Crayford, Brent South, and Brent East. In addition, it revealed the marginal constituencies of the inner Victorian suburbs like Lewisham West, Norwood, Putney, and Ealing North. But the revision of boundaries and the introduction of single-member constituencies have also introduced new features. It has been estimated that overall the re-drawing of boundaries may have cost the Labour Party about five seats but it has given them a chance to make inroads into some normally Conservative suburban areas like Mitcham and Morden and Ilford North. Moreover, in single-member seats third parties not only have a better chance of winning, as the Liberals proved in Sutton and Cheam, and in Richmond, but they can also considerably reduce the majority of the large parties to a level where the seat becomes marginal, as in Newham North East, or it actually changes hands, as in Hendon North.

1964
1967
Percentage share of vote
Conservative vote
75·0 or more
65·0–74·9
55·0–64·9
45·0–54·9
0·0–44·9
Labour vote
0·0–44·9
45·0–54·9
55·0–64·9
65·0–74·9
75·0 or more
0 1 2 3 4 5 miles
1970
1973
LIB
LIB

Afterword by Emrys Jones

Making an atlas of a city like London is rather like dissecting an animal. We are all familiar with the animal, but we want to know what makes it tick, or at least we want to know how it is built and how different parts of the body carry out different functions. So we break a city down into separate systems: the environmental shell, occupations, movement—much as the anatomist would describe the skeletal system, digestive organs, or blood system. The analogy must not be pushed too far, except for one small reminder; the animal is killed before it can be anatomically dissected, and in a slightly parallel way we have held the city still in order to show what its component parts are; it is held still on the night of the census on which many of the maps are based. This effect, which sometimes distorts at least some aspects of city life, can be overcome in some degree by measuring changes over time and movements in space; the static structure is only a departing point for an analysis of a living city.

Can we now do what the anatomist cannot do—put all the parts together and breath life into it again? Perhaps we can go some way towards this by seeing first how successfully the social life of London has been revealed in its spatial aspects and, secondly, whether they form a coherent picture which deepens our understanding of the city, if not indeed of all metropolitan cities of this kind. Let us remind ourselves that we are interested in the social geography. There have been monumental works in the past which tried to describe London's people; the two classic works are Mayhew's in the middle of the nineteenth century and Booth's at the end of the century. Both were extremely searching studies and gave a vivid picture of life in London, but neither was particularly interested in distributions and their significance. Only recently have the data been mapped in sufficient detail to reveal the fascinating mosaic of social groups in London, as complete and objective a survey as one could wish for. As geographers our attention at once swings away from the study of the social group as such to its distribution, the way in which it interacts with certain environments, its relationship to other groups, the part it plays in the total picture.

The introduction suggested a model which simplified the social aspects of London's geography in terms of four structure maps. These can be used as a framework for summing up; for they were meant to suggest the integration of the human aspects of the city with the urban environment. It seems rather obvious to state that the two major systems are inextricably interrelated; no one wishes to suggest that one 'causes' the other. But, of the variables that shape the social distributions, the built environment is one of the key ones. A city's housing is 'given', most of it set in a historic mould, reflecting past social geographies, and it is today nothing more than a shell through which social life passes. But its age, nearness to the centre, state of upkeep—all these continue to make distinctive environments which our value-system fits on to a scale of desirability. This is why so many social variables correlate; social class and housing class become the same. However much individuals may vary, and however we as individuals may exercise choice, our behaviour in apportioning our budget, in giving priority to nearness to work, and so on will conform broadly to our resources. Together these make an identifiable pattern, or life-style, and once this is linked to housing, people with the same life-style find themselves bounded by very considerable constraints; for instance, the availability of a house of the right size at a price they can afford in a part of London which they like, may be very limited. Moreover, they are not likely to have enough information to widen their choice. London is such a large city that none of us will pretend to 'know' more than a sector of it. It is not in the least surprising that the easily identifiable owner-occupier suburbs, the inner rooming sectors, the rented family accommodation, are all firmly linked with housing stock and the way in which our society appraises it.

Where people live, and whether they own or rent a house or flat, is only one aspect, on the face of it a static one. Again, we need to be reminded that the picture has been caught at one moment of time by the census. People do move. What the French geographers call circulation—movement to work, to the shops, to leisure activities—makes it clear that transport is another key element in the city. The population and density maps, with their empty core and well peopled residential and suburban areas, are turned inside out during the day, when over a million people come into central London to work, and many others to shop. The opportunities of city life and their availability are not measured by nearness only, but by accessibility, and distance from the centre must be set against the ability to pay to overcome this distance. One can live on the outskirts of the city and yet be very involved in it, an integral part of all aspects of its life. On the other hand, one can live very near the centre without the ability to profit from this nearness—in the city but not of it. The difference in life-styles, measured by constraints, is a function, more often than not, of resources—money!

The fact that very similar patterns tend to emerge for every variable, suggests that London has a kind of 'regional' pattern to its social geography. The same areas have low scores on those variables which show desirable aspects of city life—pleasant surroundings and low density, etc.—and high scores on the undesirable aspects like overcrowding and sharing facilities. Access to opportunities also conforms to a scale. There is no doubt, for example, that areas like Notting Hill, Bethnal Green, North Southwark, and Islington, are all regions of multiple deprivation. On the face ot it, there is nothing new in this statement, but the atlas shows, unemotionally and with no prejudice, exactly what makes up this deprivation in terms of the environment and the society. It is difficult to see how social policy, based on a more complete understanding of all the variables, can ignore these spatial aspects and their implications.

It is tempting to work upon immigrant groups—particularly if they have different racial characteristics—in a rather special way. The atlas has shown that, in the absence of the extreme concentration which typifies the ghetto, these groups do not behave, statistically, very differently from the socially-deprived of indigenous origin. The fact that they are distinguished in the census data makes it easier to plot their distribution and their movement but, as far as spatial relationships in a city are concerned, they act very much like poorer whites. Environmental and social poverty are city problems in which ethnic differences are incidental.

Finally, from our urban land value model, a very generalized picture of of London has emerged, and we can question the three simple patterns which were suggested in the introduction. A model can be the link between London and other major cities of the western world. Of course, there is a sense in which London is unique—in its site, its history, and its development. But it is also the outcome of western society, largely the creation of nineteenth-century industrialization and its twentieth-century aftermath. This it shares with many cities, and it is the product of such forces that give us a model we can apply widely. Obviously the contrast between inner London and outer is paramount, sometimes abrupt, sometimes a gradual merging. Environmentally, it is partly a function of age, and, as a corollary, of the adaptability of older buildings to modern life. Socially, we have already seen how closely society and environment are interwoven. Deprivation, with age and out-of-dateness, polarizes towards the centre, in the zone which Burgess called 'transitional', immediately beyond the newest high-rise city office blocks. This simple dichotomy and zoning is broken by the tendency of certain sectors, extending from near the centre to the periphery, to maintain certain life-styles for reasons which are unique to London. One clear example is the extension of the 'working class' along the river to the east; another is the extension of the traditional west end in a north-westerly direction. We can carry our generalized model no further than to say that such sectors exist and that those specific ones are unique to London: that they break up a simple basic pattern is a reminder that the urban environment is invested with social values, and that our models can only hope to accommodate the most general and clear-cut of these. This leaves—perhaps fortunately—much of the unexpectedness and mystery of city life to be learnt at first hand, part of the ambience of the city that no map can capture.

Appendices

Appendix 1

Mapping social data

The commentaries in this atlas are designed to help in the interpretation of the individual maps. But, in order to understand and appreciate the information contained in a map, it is necessary to know how it was constructed and to understand the influence of changes of *scale* on the information it portrays. Most of our maps are *choropleth* maps. In this appendix, the use of choropleth mapping is explained with reference to a specific problem—the mapping of population density within the Borough of Camden.

If there are, say, 80 people housed on a small block covering 2 hectares, we may speak of a *population density* of this area of 40 people per hectare.[1] This figure tells us the *average* number of people living in a specified area of land. But it is only a generalization—for one half of the block may consist of households of 1 and 2 persons, while the other half may have large households living at a higher density in relatively overcrowded conditions. The main factors which influence urban population density are: the size and density of residential buildings; their level of occupancy; the amount of open space around buildings (such as the size of their gardens); and the amount of other non-residential space in the area. This last factor incorporates parks, roads, and other open spaces, as well as land used for industrial and commercial purposes.

The average figure for the block could be mapped and the area could be shaded on a map in a specified way to represent its population density. Shading according to the numerical value of some form of statistical data is known as *choropleth mapping*. But there cannot be a unique shading for every specific value to be mapped, so these values must be grouped into convenient *class intervals.* In our case the block may be grouped with population densities of less than 50 persons per hectare and shaded with light and widely-spaced vertical lines to represent relatively low population densities.

Our next hypothetical block may consist simply of a 20 storey high-rise block of council flats in which some 960 people are housed on a three hectare site. This represents an average population density of 320 persons per hectare. If densities of over 300 persons per hectare are classified as the group of highest density, this block may be shaded completely black on the map; it is a fundamental convention of choropleth mapping that degrees of density, intensity, or concentration of a phenomenon are displayed by a series of shadings, graded from solid for high degrees to a very light shade for low degrees.

We could calculate the population density of each block or street in London and display these values on one map, but it would be an immense task to map such a large number of very small areas. We must, therefore, introduce another concept of grouping— in this case, amalgamating small areas into larger units which can (because of their lesser numbers) be more easily mapped. We thereby change the scale of the areas we are dealing with and, as the scale of our analysis changes, so do the average values for our small units. For example, if the two blocks considered above were amalgamated and mapped as one unit, it would have a population density of 208 persons per hectare. As the area is enlarged and the mapping is reduced, the population density figure mapped becomes more generalized. This represents a loss of information, but is compensated for by the greater ease of mapping information and of interpreting the pattern revealed by the map.

The patterns of population density which can be observed for areal units of different sizes within the borough of Camden (Maps a.—c.) clearly illustrate the influence of changes of scale and the features of choropleth mapping. In this context, a preliminary comment on the structure of London in terms of its administration and census enumeration units, is relevant. Greater London may be thought of as a nested hierarchy of these areas. Within Greater London there are 32 *boroughs*, plus the City of London. Each borough is subdivided into a number of *wards* (average—20), and in 1971 there were 654 wards in the 32 boroughs. In turn, each of the wards comprises a number of *enumeration districts* (e.d.s), which are the smallest official units of enumeration for census data and are based on the number of *households* an individual census enumerator can visit on the night of the census.

The street plan of just one of the 30 e.d.s within St. John's ward in the borough of Camden is shown in Map a. opposite. Detailed examination of this map reveals that there is a variety of land-uses in the area and that the amount of non-residential land varies considerably from street block to street block. It is almost certain that the population density for the block which contains at least three factories or warehouses is less than for other, primarily residential blocks. Yet an average population density for the entire e.d. has been calculated and Map b. (population density of all e.d.s within St. John's ward) shows that it falls into the 30—50 people per hectare class interval. In comparison to most other e.d.s in this ward

[1] One hectare = 2·47 acres.

(d) A census enumeration district in St. Johns Ward, Camden, 1971

(c) Population density of enumeration districts in St. Johns Ward, Camden, 1971

(a) Borough population density, 1971

(b) Population density of wards in Camden, 1971

this value is low; and there is considerable variation between the density of individual e.d.s. Because of the *change of scale*, it is difficult to state the specific reasons for these variations.

At the next scale of aggregation, when these individual e.d.s are considered as just one *ward* (Map c) we can see that there has been considerable loss of information. The variation within the ward has been suppressed and the average population density for the ward is shown. Similarly, variations at the ward level are suppressed when all of Camden's wards are considered as one borough. For example, the low population densities in the northern wards of the borough, which incorporate the Parliament Hill—Hampstead Heath area, are submerged by the much higher population densities over the rest of the borough; so in Map d. a single population density value for the entire borough of Camden has been mapped. Camden is seen in Map d. as a borough of relatively high population density (90—110 persons per hectare) although several of the individual wards had population densities well over 110 persons per hectare. Of course, the exercise need not stop at this point; the population density for Greater London as a whole can be calculated and compared to other regions; and the regions can be amalgamated to give a national figure.

This case study of population density within Camden illustrates the effects of changes of scale on data, and the loss of information which results from amalgamating small areal units. At all scales, the average figure for the areal units and data being considered are conveyed visually by the intensity of the shading used on the choropleth map.

Appendix 2

The components of pattern

The maps of education, health and environment, and revenue expenditure in the London boroughs were prepared with the aid of a statistical technique known as *principal components analysis.* The main purpose of this technique is to bring out the underlying similarities or patterns in a table of statistical information. It is possible to do this because many of the characteristics of society that we measure have been found to be closely related or *correlated*. We have already made visual correlations between several maps in this atlas by noting that areas which have, for example, high rates of owner-occupancy, also tend to have high car-ownership and high proportions of middle-class, Conservative-voting residents. Principal components analysis is used to bring together distinct groups of correlated characteristics or *variables* and combine them into significantly fewer groups or *components* each of which serves as a composite measure but which are themselves uncorrelated.

There are many variables which relate to education and these can be obtained from official sources and mapped. But which ones are most important and will tell us most about the general character of education in London? Our own attempts to select the more broadly relevant variables still left us with a total of thirteen. Clearly, mapping all these variables separately would have involved considerable repetition, since the underlying pattern of one map would recur on maps of other, correlated variables.

Correlations between variables can be either *direct* or *indirect*. In the case of direct correlation, high values of one variable generally correspond to high values of another. If there is indirect correlation between two variables, then high values for one of them corresponds to low values for the other. In other words, correlation is a method of matching the extent and the direction of the variation or *variability* in variables. This appendix explains how we arrived at the components for education; the method was also used for the subsequent maps in Section 7.

Four variables are involved in the first component of education:- the proportion of eighteen and nineteen year-olds in a borough receiving full and lesser-value awards from their local authority for university study; the proportion entering colleges of education; children at school aged sixteen as a proportion of those aged thirteen three years earlier; and the ratio of male to female students receiving full-value grants to go to university. When four separate maps were drawn for each of these variables the overall patterns were very similar. The essential contrast was between inner and outer London. Suburban boroughs to the north-west and south like Bromley, Croydon, Merton, Harrow, and Barnet, gave proportionately more grants for university study, had more students going to colleges of education, and had higher rates of staying on at school. Inner London boroughs generally and Newham, Barking, and Bexley in particular, had low values on these variables. These three variables were therefore all strongly and directly correlated. Boroughs granting proportionately more awards were also fairer in distributing them between the sexes. Bromley and the others mentioned above had low ratios of male to females receiving awards, while the boroughs which gave fewer awards overall also gave many more to men than to women. In terms of the maps, this variable had the same general pattern as the others but where they were shaded for high values on the 'awards' variables this one was shaded for low values on male:female ratios. It was therefore indirectly correlated with 'awards and staying at school.'

It is because these four variables have such high intercorrelations that they form the first component which can be given a name to suggest the nature of the most important variables incorporated in it. The component highlights the association between the granting of awards and the tendency for children to stay on at school and we can therefore label it *awards and staying at school*. The proportion of the total variation in the original data accounted for by each component can be calculated. The first component is always the most important underlying pattern in the original information and, in our case, 'the awards and staying at school' component accounts for 32 per cent of the total variation and subsequent components account for lesser proportions of total variability.

The components can now be related back to the areas of observation (London boroughs) and a numerical value assigned to them by calculating a composite index for every borough, consisting of the values it had for the original variables multiplied by the extent to which those variables are involved in the new component. It is these *component scores* which are then displayed in map form. In the commentaries to the component maps we usually refer to a component by its general name, that is treating it as a variable in its own right, but the keys to the maps and the detailed list of variables make clear the relationship between the underlying variables in each component.

Having found the first component by discovering the group of original variables that are most highly correlated, the second component is similarly defined as a group of variables which are correlated among themselves but which, as a group, are totally *unrelated* to the first component. Subsequent components are similarly unrelated to previous components which means that the index scores calculated for the remaining components are also uncorrelated, a property which is very useful for the interpretation of components maps since the components will usually convey strongly contrasting map patterns.

The second education component has been called *size and denomination of primary schools*. Four variables relating to primary education are involved in it. One end of the spectrum distinguishes boroughs with large primary schools and high pupil:teacher ratios; the opposite end shows boroughs with large proportions of children in denominational schools (Church of England and Roman Catholic). Boroughs with bigger schools also appear to have fewer denominational schools.

The third component is called *selectivity and immigrants in secondary education* and it contrasts those boroughs with high proportions of children still in grammar, technical, and modern schools against those with more immigrant school-children in comprehensive schools. These two components accounted for a further 23·4 per cent and 13·3 per cent respectively of the total variation in the original information. So, with just three maps based on scores derived from these components, we can convey roughly 70 per cent of the information contained in the original thirteen variables.

Variables and components of pattern

1. Education provision

Variable	Relation to component D(irect)/I(ndirect)	Component	Variation accounted for (per cent)
i Entrants to Colleges of Education, 1971	D	I 'Awards and staying at school'	31·8
ii Full and lesser value awards at universities per 1000 of 18-19 age group, 1971	D		
iii Pupils aged 16 as percentage of those aged 13 three years previously, 1971	D		
iv Ratio of men:women receiving full value awards, 1971	I		
i Average number of full-time pupils in primary schools, 1972	D	II 'Primary schools'	23·4
ii Pupil:teacher ratio in primary schools, 1972	D		
iii Percentage of primary school pupils in C. of E. schools, 1972	I		
iv Percentage of primary school pupils in R.C. schools, 1972	I		
i Percentage of secondary school pupils in modern schools, 1972	D	III 'Secondary schools'	13·3
ii Percentage of secondary pupils in grammar and technical schools, 1972	D		
iii Percentage of secondary pupils in fully comprehensive schools, 1972.	I		
iv Percentage immigrant pupils	I		
i Percentage of secondary pupils in other secondary schools	Not involved in first three components.		

2. Metropolitan environment

Variable	Relation to component D(irect)/I(ndirect)	Component	Variation accounted for (per cent)
i Five year average of persons killed in road accidents per road mile, 1967–71	D	I 'Inner city environment'	40·0
ii Population density, 1971.	D		
iii Percentage of people referred as mentally-ill or handicapped, 1971	D		
iv Mileage of major roads per acre of residential land-use, 1966	D		
v Five-year average of smoke particles per cubic metre of air, 1967–71	D		
vi Five-year average SO_2 per cubic metre of air, 1967–71	D		
vii Illegitimate births per 1000 births, 1971	D		
i Death rate from heart and circulatory diseases, 1971	D	II 'Deaths from circulatory disease and certain cancers'	14·6
ii Death rate from cancer other than lung and bronchus, 1971	D		
iii Acres of open space per head of population	D		

Variable	Loading	Component	% variance
i Still births per 1000 live births, 1971	D	III 'Air pollution, bad living conditions and respiratory diseases'	14·1
ii Deaths from respiratory diseases, 1971	D		
iii Percentage of dwellings in 'poor or unfit' condition, 1967	D		
iv Five-year average of smoke particles per cubic metre of air	D		
v Five-year average SO_2 per cubic metre of air	D		
i Death rates from cancer of the lung and bronchus		Not involved in first three components.	
ii Percentage of children failing school medical test			

Revenue Expenditure 1970–71

Variable	Loading	Component	% variance
i Education and youth employment	D	I 'Education, health, and welfare'	54·0
ii Public libraries	D		
iii Personal health services	D		
iv Welfare services	D		
v Child care	D		
vi Public baths	D		
vii Public health	D		
i Town and Country Planning	D	II 'Planning and open space'	16·0
ii Paths and open spaces	D		
i Road safety		Not involved in first two components	

Appendix 3 Key to London Wards, 1971

Barking
1 Abbey
2 Cambell
3 Chadwell Heath
4 Eastbrook
5 Fanshawe
6 Gascoigne
7 Heath
8 Longbridge
9 Manor
10 River
11 Valence
12 Village

Barnet
1 Arkley
2 Brunswick Park
3 Burnt Oak
4 Childs Hill
5 Colindale
6 East Barnet
7 East Finchley
8 Edgware
9 Finchley
10 Friern Barnet
11 Garden Suburb
12 Golders Green
13 Hadley
14 Hale
15 Hendon
16 Mill Hill
17 St. Paul's
18 Totteridge
19 West Hendon
20 Woodhouse

Bexley
1 Belvedere
2 Bostall
3 Brampton
4 Christ Church
5 Crayford North
6 Crayford Town
7 Crayford West
8 Danson
9 East Wickham
10 Erith Town
11 Falconwood
12 Lamorbey East
13 Lamorbey West
14 North Cray
15 Northumberland Heath
16 St. Mary's
17 St. Michael's
18 Sidcup East
19 Sidcup West
20 Upton

Brent
1 Alperton
2 Barham
3 Brentwater
4 Brondesbury Park
5 Carlton
6 Chamberlayne
7 Church End
8 Cricklewood
9 Fryent
10 Gladstone
11 Harlesden
12 Kensal Rise
13 Kenton
14 Kilburn
15 Kingsbury
16 Manor
17 Mapesbury
18 Preston
19 Queensbury
20 Queen's Park
21 Roe Green
22 Roundwood
23 St. Raphael's
24 Stonebridge
25 Sudbury
26 Sudbury Court
27 Tokyngton
28 Town Hall
29 Wembley Central
30 Wembley Park
31 Willesden Green

Bromley
1 Anerley
2 Bickley
3 Biggin Hill
4 Bromley Common
5 Chelsfield
6 Chislehurst
7 Clock House
8 Copers Cope
9 Darwin
10 Eden Park
11 Farnborough
12 Goddington
13 Keston and Hayes
14 Lawrie Park and Kent House
15 Manor House
16 Martin's Hill and Town
17 Mottingham
18 Penge
19 Petts Wood
20 Plaistow and Sundridge
21 St. Mary Cray
22 St. Paul's Cray
23 Shortlands
24 West Wickham North
25 West Wickham South

Camden
1 Adelaide
2 Belsize
3 Bloomsbury
4 Camden
5 Chalk Farm
6 Euston
7 Gospel Oak
8 Grafton
9 Hampstead Central
10 Hampstead Town
11 Highgate
12 Holborn
13 Kilburn
14 Kings Cross
15 Priory
16 Regent's Park
17 St. John's
18 St. Pancras
19 West End

Croydon
1 Addiscombe
2 Bensham Manor
3 Broad Green
4 Central
5 Coulsdon East
6 East
7 New Addington
8 Norbury
9 Purley
10 Sanderstead and Selsdon
11 Sanderstead North
12 Shirley
13 South Norwood
14 Thornton Heath
15 Upper Norwood
16 Waddon
17 West Thornton
18 Whitehorse Manor
19 Woodcote and Coulsdon West
20 Woodside

Ealing
1 Brent
2 Central
3 Cleveland
4 Dormers Wells
5 East
6 Elthorne
7 Glebe
8 Hanger Hill
9 Heathfield
10 Horsenden
11 Mandeville
12 Northcote
13 Northfields
14 Perivale
15 Ravenor
16 Southfield
17 Springfield
18 Walpole
19 Waxlow Manor
20 West End

Enfield
1 Angel Road
2 Arnos
3 Bowes
4 Bullsmoor
5 Bush Hill
6 Bush Hill South
7 Cambridge Road
8 Chase
9 Church Street
10 Cockfosters
11 Craig Park
12 Enfield Wash
13 Grange
14 Green Street
15 Highfield
16 Jubilee
17 New Park
18 Oakwood
19 Ordnance
20 Palmers Green
21 Ponders End
22 Pymmes
23 St. Alphege
24 St. Peter's
25 Silver Street
26 Southgate Green
27 Town
28 West
29 Willow
30 Winchmore Hill

Greenwich
1 Abbey Wood
2 Academy
3 Blackheath
4 Charlton
5 Coldharbour
6 Eastcombe
7 Eltham
8 Eynsham
9 Hornfair
10 Horn Park
11 Kidbrooke
12 Marsh
13 Middle Park
14 New Eltham
15 Park
16 St. George's
17 St. Margaret's
18 St. Mary's
19 St. Nicholas
20 Sherard
21 Shooters Hill
22 Slade
23 Trafalgar
24 Vanbrugh
25 Well Hall
26 West
27 Woolwich

Hackney
1 Brownswood
2 Chatham
3 Clissold
4 Dalston
5 De Beauvoir
6 Defoe
7 Downs
8 Haggerston
9 Kingsmead
10 Leabridge
11 Moorfields
12 New River
13 Northfield
14 Northwold
15 Queensbridge
16 Rectory
17 Springfield
18 Victoria
19 Wenlock
20 Wick

Hammersmith
1 Addison
2 Avonmore
3 Broadway
4 Brook Green
5 Colehill
6 College Park and Old Oak
7 Coningham
8 Crabtree
9 Gibbs Green
10 Grove
11 Halford
12 Margravine
13 Parson's Green
14 St. Stephens
15 Sandford
16 Sherbrooke
17 Starch Green
18 Sulivan
19 Town
20 White City
21 Wormholt

Haringey
1 Alexandra-Bowes
2 Bruce Grove
3 Central Hornsey
4 Coleraine
5 Crouch End
6 Fortis Green
7 Green Lanes
8 High Cross
9 Highgate
10 Muswell Hill
11 Noel Park
12 Park
13 Seven Sisters
14 South Hornsey
15 South Tottenham
16 Stroud Green
17 Tottenham Central
18 Town Hall
19 Turnpike
20 West Green

Harrow
1 Belmont
2 Harrow-on-the-Hill and Greenhill
3 Harrow Weald
4 Headstone
5 Kenton
6 Pinner North and Hatch End
7 Pinner South
8 Queensbury
9 Roxbourne
10 Roxeth
11 Stanmore North
12 Stanmore South
13 Wealdstone North
14 Wealdstone South
15 West Harrow

Havering
1 Bedfords
2 Central
3 Collier Row
4 Cranham
5 Elm Park
6 Emerson Park
7 Gidea Park
8 Gooshays
9 Hacton
10 Harold Wood
11 Heath Park
12 Heaton
13 Hilldene
14 Hylands
15 Mawney
16 Oldchurch
17 Rainham
18 St. Andrew's
19 South Hornchurch
20 Upminster

Hillingdon
1 Belmore
2 Colham-Cowley
3 Eastcote
4 Frogmore
5 Harefield
6 Haydon
7 Hayes
8 Hillingdon East
9 Hillingdon West
10 Ickenham
11 Manor
12 Northwood
13 Ruislip
14 South
15 South Ruislip
16 Uxbridge
17 Yeading
18 Yiewsley

Hounslow
1 Clifden
2 Cranford
3 East Bedfont
4 Feltham Central
5 Feltham North
6 Feltham South
7 Gunnersbury
8 Hanworth
9 Heston East
10 Heston West
11 Homefields
12 Hounslow Central
13 Hounslow Heath
14 Hounslow South
15 Hounslow West
16 Isleworth North
17 Isleworth South
18 Riverside
19 Spring Grove
20 Turnham Green

Islington
1 Barnsbury
2 Bunhill
3 Canonbury
4 Clerkenwell
5 Highbury
6 Highview
7 Hillmarton
8 Hillrise
9 Holloway
10 Junction
11 Mildmay
12 Parkway
13 Pentonville
14 Quadrant
15 St. George's
16 St. Mary
17 St. Peter
18 Station
19 Thornhill

Kensington & Chelsea
1 Brompton
2 Cheyne
3 Church
4 Earl's Court
5 Golborne
6 Hans Town
7 Holland
8 Norland
9 North Stanley
10 Pembridge
11 Queen's Gate
12 Redcliffe
13 Royal Hospital
14 St. Charles
15 South Stanley

Kingston upon Thames
1 Berrylands
2 Burlington
3 Cambridge
4 Canbury
5 Chessington
6 Coombe
7 Dickerage
8 Grove
9 Hill

(continued overleaf)

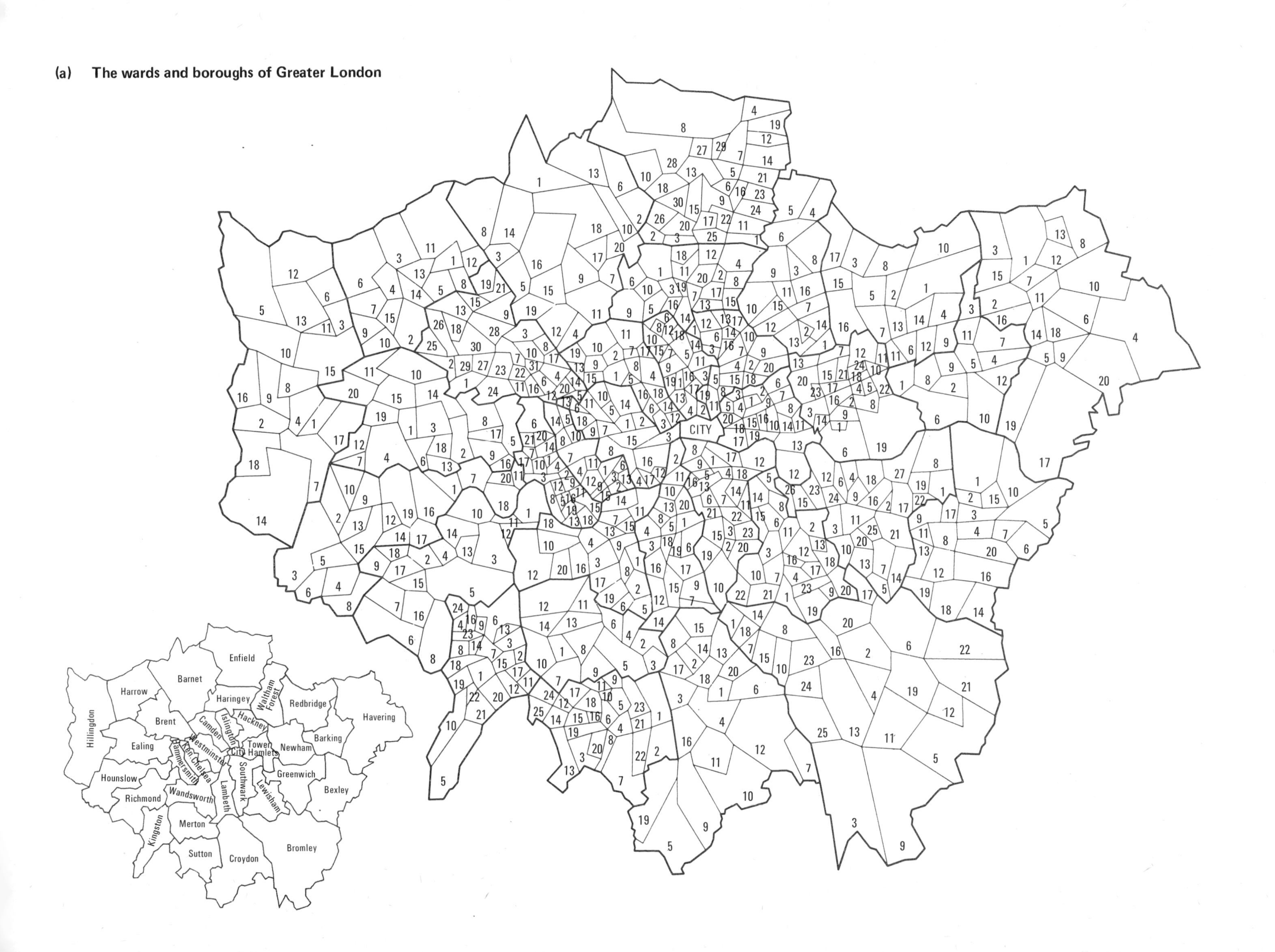

(a) **The wards and boroughs of Greater London**

10 Hook and
Southborough
11 Malden Green
12 Malden Manor
13 Mount
14 Norbiton
15 Norbiton Park
16 Park
17 St. James's
18 St. Mark's and
Seething Wells
19 Surbiton Hill
20 Tolworth East
21 Tolworth South
22 Tolworth West
23 Town
24 Tudor

Lambeth
1 Angell
2 Bishop's
3 Clapham Park
4 Clapham Town
5 Ferndale
6 Herne Hill
7 Knight's Hill
8 Larkhall
9 Leigham
10 Oval
11 Prince's
12 St. Leonard's
13 Stockwell
14 Streatham South
15 Streatham Wells
16 Thornton
17 Thurlow Park
18 Town Hall
19 Tulse Hill
20 Vassall

Lewisham
1 Bellingham
2 Blackheath and
Lewisham Village
3 Brockley
4 Culverley
5 Deptford
6 Drake
7 Forest Hill
8 Grinling Gibbons
9 Grove Park
10 Honor Oak Park
11 Ladywell
12 Lewisham Park
13 Manor Lee
14 Marlowe
15 Pepys
16 Rushey Green
17 St. Andrew
18 St. Mildred Lee
19 Southend
20 South Lee
21 Sydenham East
22 Sydenham West
23 Whitefoot

Merton
1 Cannon Hill
2 Mitcham Central
3 Mitcham East
4 Mitcham North
5 Mitcham South
6 Mitcham West
7 Morden
8 Priory
9 Ravensbury
10 West Barnes
11 Wimbledon East
12 Wimbledon North
13 Wimbledon South
14 Wimbledon West

Newham
1 Beckton
2 Bemersyde
3 Canning Town and
Grange
4 Castle
5 Central
6 Custom House and
Silvertown
7 Forest Gàte
8 Greatfield
9 Hudsons
10 Kensington
11 Little Ilford
12 Manor Park
13 New Town
14 Ordnance
15 Park
16 Plaistow
17 Plashet
18 St. Stephens
19 South
20 Stratford
21 Upton
22 Wall End
23 West Ham
24 Woodgrange

Redbridge
1 Aldborough
2 Barkingside
3 Bridge
4 Chadwell
5 Clayhall
6 Clementswood
7 Cranbrook
8 Fairlop
9 Goodmayes
10 Hainault
11 Ilford
12 Mayfield
13 Park
14 Seven Kings
15 Snaresbrook
16 Wanstead
17 Woodford

Richmond upon Thames
1 Barnes
2 Central Twickenham
3 East Sheen
4 East Twickenham
5 Ham/Petersham
6 Hampton
7 Hampton Hill
8 Hampton Wick
9 Heathfield
10 Kew
11 Mortlake
12 Palewell
13 Richmond Hill
14 Richmond Town
15 South Twickenham
16 Teddington
17 West Twickenham
18 Whitton

Southwark
1 Abbey
2 Alleyn
3 Bellenden
4 Bricklayers
5 Browning
6 Brunswick
7 Burgess
8 Cathedral
9 Chaucer
10 College
11 Consort
12 Dockyard
13 Faraday
14 Friary
15 Lyndhurst
16 Newington
17 Riverside
18 Rotherhithe
19 Ruskin
20 Rye
21 St. Giles
22 The Lane
23 Waverley

Sutton
1 Beddington North
2 Beddington South
3 Belmont
4 Carshalton Central
5 Carshalton North East
6 Carshalton North West
7 Carshalton South East
8 Carshalton South West
9 Carshalton St. Helier
North
10 Carshalton St. Helier
South
11 Carshalton St. Helier
West
12 Cheam North
13 Cheam South
14 Cheam West
15 Sutton Central
16 Sutoon East
17 Sutton North
18 Sutton North East
19 Sutton South
20 Sutton South East
21 Wallington Central
22 Wallington North
23 Wallington South
24 Worcester Park North
25 Worcester Park South

Tower Hamlets
1 Bethnal Green Central
2 Bethnal Green East
3 Bethnal Green North
4 Bethnal Green South
5 Bethnal Green West
6 Bow North
7 Bow South
8 Bromley
9 Holy Trinity
10 Limehouse
11 Poplar East
12 Poplar Millwall
13 Poplar South
14 Poplar West
15 Redcoat
16 St. Dunstan's
17 St. Katharine's
18 St. Mary's
19 Shadwell
20 Spitalfields

Waltham Forest
1 Cann Hall
2 Central
3 Chapel End
4 Chingford Central
5 Chingford North West
6 Chingford South
7 Forest
8 Hale End
9 Higham Hill
10 High Street
11 Hoe Street
12 Lea Bridge
13 Leyton
14 Leytonstone
15 St. James Street
16 Wood Street

Wandsworth
1 Balham
2 Bedford
3 Earlsfield
4 Fairfield
5 Furzedown
6 Graveney
7 Latchmere
8 Nightingale
9 Northcote
10 Putney
11 Queenstown
12 Roehampton
13 St. John
14 St. Mary's Park
15 Shaftesbury
16 Southfield
17 Springfield
18 Thamesfield
19 Tooting
20 West Hill

Westminster, City of
1 Baker Street
2 Cavendish
3 Charing Cross
4 Churchill
5 Church Street
6 Harrow Road
7 Hyde Park
8 Knightsbridge
9 Lancaster Gate
10 Lords
11 Maida Vale
12 Millbank
13 Queen's Park
14 Regent's Park
15 Regent Street
16 Victoria Street
17 Warwick
18 Westbourne

Key to London Parliamentary Constituencies, 1971

Barking
1 Barking
2 Dagenham

Barnet
3 Chipping Barnet
4 Finchley
5 Hendon North
6 Hendon South

Bexley
7 Bexleyheath
8 Erith and Crayford
9 Sidcup

Brent
10 Brent East
11 Brent North
12 Brent South

Bromley
13 Beckenham
14 Chislehurst
15 Orpington
16 Ravensbourne

Camden
17 Hampstead
18 Holborn and
19 St. Pancras South

Croydon
20 Croydon Central
21 Croydon North East
22 Croydon North West
23 Croydon South

Ealing
24 Acton
25 Ealing North
26 Southall

Enfield
27 Edmonton
28 Enfield North
29 Southgate

Greenwich
30 Greenwich
31 Woolwich East
32 Woolwich West

Hackney
33 Hackney Central
34 Hackney North and
Stoke Newington
35 Hackney South and
Shoreditch

Hammersmith
36 Fulham
37 Hammersmith North

Haringey
38 Hornsey
39 Tottenham
40 Wood Green

Harrow
41 Harrow Central
42 Harrow East
43 Harrow West

Havering
44 Hornchurch
45 Romford
46 Upminster

Hillingdon
47 Hayes and Harlington
48 Ruislip-Northwood
49 Uxbridge

Hounslow
50 Brentford and
Isleworth
51 Feltham and Heston

Islington
52 Islington Central
53 Islington North
54 Islington South and
Finsbury

Kensington & Chelsea
55 Chelsea
56 Kensington

Kingston upon Thames
57 Kingston upon Thames
58 Surbiton

Lambeth
59 Lambeth Central
60 Norwood
61 Streatham
62 Vauxhall

Lewisham
63 Deptford
64 Lewisham East
65 Lewisham West

Merton
66 Mitcham and Morden
67 Wimbledon

Newham
68 Newham North East
69 Newham North West
70 Newham South

Redbridge
71 Ilford North
72 Ilford South
73 Wanstead and
Woodford

Richmond upon Tham
74 Richmond
75 Twickenham

Southwark
76 Bermondsey
77 Dulwich
78 Peckham

Sutton
79 Carshalton
80 Sutton and Cheam

Tower Hamlets
81 Bethnal Green and
Bow
82 Stepney and Poplar

Waltham Forest
83 Chingford
84 Leyton
85 Walthamstow

Wandsworth
86 Battersea North
87 Battersea South
88 Putney
89 Tooting

Westminster, City of
90 Part of The City of
London and West-
minster South
91 Paddington
92 St. Marylebone